Gregory L. Moss
Purdue University

D1198449

Lab Manual (A Design Approach)

to accompany

DIGITAL SYSTEMS
Principles and Applications

Seventh Edition

Ronald J. Tocci
Neal S. Widmer

Prentice Hall
Upper Saddle River, New Jersey   Columbus, Ohio

Editor: Linda R. Ludewig
Developmental Editor: Carol Hinklin Robison
Production Editor: Mary Harlan
Cover Designer: Brian Deep

This book was printed and bound by Courier/Kendallville, Inc. The cover was printed by Phoenix Color Corp.

 © 1998 by Prentice-Hall, Inc.
Simon & Schuster/A Viacom Company
Upper Saddle River, New Jersey 07458

Printed in the United States of America

10 9 8 7 6 5 4 3 2 1

ISBN: 0-13-745894-0

Prentice-Hall International (UK) Limited, *London*
Prentice-Hall of Australia Pty. Limited, *Sydney*
Prentice-Hall Canada Inc., *Toronto*
Prentice-Hall Hispanoamericana, S. A., *Mexico*
Prentice-Hall of India Private Limited, *New Delhi*
Prentice-Hall of Japan, Inc., *Tokyo*
Simon & Schuster Asia Pte. Ltd., *Singapore*
Editora Prentice-Hall do Brasil, Ltda., *Rio de Janeiro*

CONTENTS

To my family,
Marita, David, and Ryan

PREFACE

This laboratory manual was written for students in an introductory digital electronics course that emphasizes logic circuit analysis, applications, and design. The approach taken in this manual is quite different from other introductory digital lab manuals. The laboratory projects are not designed to be merely "build this circuit and fill-in the blanks in the manual" but, instead, require that students "show their work" in circuit analysis and design as well as demonstrate their ability to construct and troubleshoot digital circuitry. This edition of the lab manual has been extensively revised, with several new projects added. The unit on D/A and A/D conversion has been significantly expanded. The units describing the PLD compiler software have been updated to describe the Windows version of CUPL by Logical Devices, Inc. The PLD compiler used in the laboratory projects is an educational version of this popular industry-standard software package. The major difference in the educational version is that it does not contain the full parts library. This easy to use software is available free for textbook adopters through Prentice-Hall.

This seventh edition lab manual is divided into 23 major topical units, with each unit containing a variety of laboratory projects. Each manual unit contains an introductory discussion of the digital topic addressed, and most units also provide one or more examples of procedures and applications. Both standard digital device and custom PLD designs are covered. In depth coverage and many design examples using PLDs and CUPL are provided. The sequencing of topics primarily follows the manual's accompanying text, *Digital Systems: Principles and Applications,* Seventh Edition, by Ronald Tocci and Neal Widmer. The manual, however, is designed for flexibility by including many different lab projects. It is possible to re-sequence many of the projects and topics to fit a particular course. The ratio of student laboratory projects dealing with standard digital devices and with PLDs can be easily adjusted to fit the goals of any electronics program. The manual is designed to provide an extensive selection of projects for a two semester introductory digital course sequence. Multiple projects are provided in each manual unit to allow variety and flexibility in the assigning of student laboratory experiments. The author does not expect that any course would have sufficient time for any student to perform all of the laboratory projects in most of the units. Rather, instructors may select appropriate projects to assign for their particular course needs. The laboratory assignments can be tailor-made to the specific objectives of the course. Additional projects may be selected for student enrichment activities.

The applications-oriented lab projects are designed to provide beginning electronics technology students with extensive experience in the analysis and design of digital logic circuits. The lab assignments consist of circuit projects that range from investigating basic logic concepts to synthesizing circuits for new applications. The projects are intended to challenge all students and to provide them with some directed laboratory experience that develops insight in digital principles, applications, and techniques of logic circuit analysis and design.

Personal computers and electronic design automation software are having a significant impact in industry today in the way digital systems are designed and developed. These new tools need to be included in the educational experience of future electronics personnel. An extremely important and rapidly expanding digital technology area is programmable logic devices. The custom implementation of logic circuits using programmable logic devices is included in several lab projects throughout the manual. In addition to extensive application of standard medium-scale, integrated logic devices, the manual provides comprehensive laboratory experience with standard PLD design tools used by industry for logic circuit design and development. Procedures for using the CUPL software are provided in two units of the lab manual, one for combinational circuit applications (Unit 6) and the other for sequential applications (Unit 13). This Windows based logic compiler is extremely easy to use and offers a variety of options for PLD design. Each of these units also provides several example PLD projects and solutions to assist students in developing PLD designs for the lab projects. Also included in the lab manual (Appendix B) is a discussion and examples for using the PLD logic simulation function provided by CUPL. The applications of PLDs, while very important in industrial practice today, may be treated as optional projects for digital courses that do not currently utilize computer applications and may be replaced with other standard logic circuit design projects that are also included in the manual.

A list of laboratory equipment, integrated circuits, and other necessary components is found in the Equipment List on the next page. The standard parts that are used in this manual may be obtained from many electronics suppliers. Manufacturers' data sheets for all ICs in the Equipment List are found in Appendix C of this manual.

I am very grateful to the companies who have helped to make this laboratory manual possible. In particular, I wish to thank Logical Devices and Lattice Semiconductor.

Gregory L. Moss
Purdue University

EQUIPMENT LIST

Laboratory Equipment & Software

Digital breadboarding system
Power supply (5v, 500ma)
Logic probe
Digital voltmeter
Oscilloscope (dual-trace minimum, preferably 4-trace)
Frequency counter
Signal generator
PLD (or universal) programmer
Personal computer (IBM or compatible)
CUPL by Logical Devices, Inc.

Digital Integrated Circuits

2	74LS00	Quad 2-input NAND
1	74HC00	Quad 2-input NAND (CMOS)
2	74LS02	Quad 2-input NOR
2	74LS04	Hex INVERTERS
1	74HCT05	Hex INVERTERS (Open Drain)
2	74LS08	Quad 2-input AND
2	74LS10	Triple 3-input NAND
1	74LS14	Hex Schmitt-trigger INVERTERS
2	74LS20	Dual 4-input NAND
1	74LS27	Triple 3-input NOR
1	74LS32	Quad 2-input OR
2	74LS47	BCD-to-7-segment DECODER/DRIVER
1	74LS75	4-bit bistable LATCH
1	74LS83A	4-bit binary FULL ADDER (or 74LS283)
1	74LS85	4-bit MAGNITUDE COMPARATOR
1	74LS86A	Quad 2-input EXCLUSIVE-OR
1	74LS90	Decade COUNTER
1	74LS93	4-bit binary COUNTER
2	74LS95B	4-bit, parallel-access SHIFT REGISTER
2	74LS112A	Dual JK, negative-edge triggered FLIP-FLOPS
2	74LS138	3-line-to-8-line DECODER/DEMULTIPLEXER
2	74LS148	8-line-to-3-line priority ENCODER

1	74150	1-of-16 MULTIPLEXER
1	74LS151	1-of-8 MULTIPLEXER
1	74LS160A	Synchronous decade COUNTER
1	74LS164	8-bit, serial-in, parallel-out SHIFT REGISTER
1	74LS166A	8-bit, parallel-in, serial-out SHIFT REGISTER
1	74LS190	Synchronous up/down decade COUNTER
1	74LS191	Synchronous up/down binary COUNTER
1	74LS193	Synchronous, 4-bit, up/down binary COUNTER
1	74LS221	Dual MONOSTABLE MULTIVIBRATOR
1	74LS244	Octal 3-state BUFFER
1	74S260	Dual 5-input NOR
1	74LS373	Octal D-type LATCH (PIPO REGISTER)
1	74LS393	Dual 4-bit binary COUNTER
2	2114	Static RAM (1K x 4)
1	GAL16V8	Electrically Erasable Programmable Logic Device

Linear Integrated Circuits

1	NE555	Timer
1	AD557	8-bit digital-to-analog converter
1	ADC0804	8-bit analog-to-digital converter
1	LM393	Voltage comparator

Miscellaneous Components

MAN72 (or equivalent) common-anode, 7-segment LED display (2 parts)

Resistors (1/4 watt):

330 Ω (10 parts)	1KΩ (10 parts)	2.2KΩ (2 parts)	3.3KΩ
10KΩ (2 parts)	16KΩ	24KΩ	27KΩ
33KΩ	47KΩ	62KΩ	68KΩ
82KΩ	100KΩ	270KΩ	

Capacitors:

10 μf	0.01 μf (8 parts)	0.001 μf	0.0047 μf
150 pf			

Potentiometers (10-turn):

10KΩ	50KΩ

Rectifier diode: 1N4001

Grayhill 84BB1-003 (or equivalent) Keypad (4 x 4 matrix) – optional

SPST switches, pushbuttons, and LEDs -- optional

DIGITAL TEST EQUIPMENT: INTRODUCTION

Objective

- To describe the function and operation of a typical digital breadboarding and testing system.

Suggested Parts	
74LS08	74LS32

Digital Test Equipment

Breadboarding and testing equipment used with digital circuits generally include the following functions:

Power supply

The power supply provides a regulated +5v DC voltage to be used to power TTL integrated circuits. Note that some units may also contain additional fixed voltages or a variable DC power source for other types of circuits.

Lamp monitors

The lamp monitors indicate the voltage level at various points in the digital circuit being tested. The lamp monitors will light when a digital "high" voltage is applied to them.

Logic switches

The logic switches input either of the two logic levels (voltages) to the circuit being tested. A switch in the "down" position will provide a logic "low" voltage, while a switch in the "up" position will provide a logic "high" voltage.

Pushbuttons or pulsers

A pushbutton provides a momentary "bounce free" logic input to the test circuit. Pulsers that are provided on some testing systems produce a short duration (narrow) pulse or change in the logic level of the pulser's output.

Clock

The clock provides a variable frequency pulse waveform that can be used for the timing control of some digital circuits.

Breadboarding sockets

Breadboarding sockets are very convenient devices on which circuits may be constructed for testing purposes. The socket contains a matrix of contacts that are used to interconnect the various components and wires needed to construct the digital circuit. The socket holes are small and only #30 to #22 solid "jumper" wires should be inserted into them. A typical breadboarding socket is illustrated in Fig. 1-1. This style of socket is designed to breadboard DIP (Dual-in-Line Package) type integrated circuits (ICs or chips). The socket shown has several separate electrical buses across the top and bottom of the board. Not all breadboarding sockets will have these buses or may have fewer of them. The buses will often be used to connect power and ground to **each** of the chips in the circuit. The chips will be inserted into the socket so that their pins will be parallel to and on either side of the center groove in the socket. Electrical connections are made to any pin by inserting wires into the holes that line up with that pin.

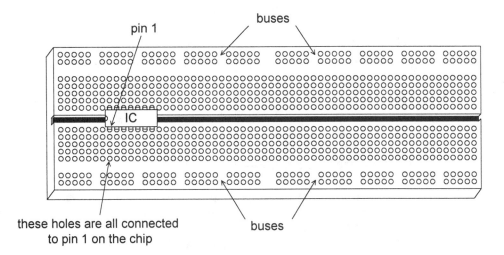

Fig. 1-1 Typical IC breadboarding socket

Test Equipment Operation

Measuring Voltage with a Digital Volt Meter (DVM)

The potential difference or voltage between the positive and negative terminals of a battery or power supply connected to a circuit will cause current to flow through the circuit. The basic unit to measure potential difference is the volt (v). A potential difference or voltage drop occurs across the various devices in a circuit when current flows through them. The magnitude (and polarity with respect to a reference point in the circuit) of a potential difference is measured with an instrument called a voltmeter. You must be extremely careful when making voltage measurements since the measurement is made on a "live" (powered) circuit. The voltmeter test leads (probes) are placed across (in parallel with) the device or power source whose voltage is to be measured. The procedure to measure DC voltages using a typical digital volt meter is:

(1)   Connect the test leads to the DVM
(2)   Set the function switch to measure DC voltages
(3)   Set the range switch for the maximum voltage anticipated
(4)   Connect (or touch) the black test lead to the reference point of the circuit (or component)
(5)   Connect (or touch) the red test lead to the point in the circuit where you wish to measure the voltage
(6)   Read the voltage on the digital display
(7)   Readjust the range setting if necessary for a proper reading

Ask your lab instructor if you have any questions concerning the use of the DVM.

Measuring Logic Levels with a Logic Probe

The logic probe is an extremely handy and easy to use piece of digital test equipment. It is used to detect and display the logic levels at various test points within a circuit.

To use a logic probe to test TTL circuits:

(1) Connect the alligator clip leads to the power supply for the circuit being tested (red to +5v and black to ground)
(2) Set the logic family switch to TTL
(3) Carefully touch the probe tip to the circuit node (normally a chip pin) to be tested (do not short any nodes together in the process)
(4) Note the logic level present at the test point by which LED (HIGH or LOW) is illuminated
(5) If neither HIGH nor LOW is indicated, the proper logic voltage is not present at the test point

Some logic probes also have a pulse detector (indicated on a separate LED) feature to indicate that the logic level at the test point is changing. Ask your lab instructor if you have any questions concerning the use of the logic probe.

Laboratory Projects

Investigate the features of the digital breadboarding and test system available in the laboratory by performing the following tasks with the unit. Carefully plug it into the AC outlet and turn on the power. Consult your lab instructor if you have any problems or questions concerning the laboratory procedures or equipment operation.

1.1 Measure and record the TTL power supply output voltage (with respect to ground) using a DVM. Remember that one of the most important safety precautions to observe in electronics is to avoid personal contact with any voltage source or component in a "live" circuit. You may need a short piece of jumper wire if the DVM's probes do not easily make electrical contact to the power supply connectors. Make sure that you are reading the TTL supply voltage if the unit has more than the single output supply. The TTL supply voltage should be between +4.75v and +5.25v DC. If you are unfamiliar with the use of the DVM, refer to the "Measuring Voltage with a Digital Volt Meter (DVM)" section of this lab assignment.

1.2 Locate the logic switches on the digital tester and determine the number of individual input switches available. Measure and record the voltage (with respect to ground) from a logic switch when it is placed in each of its two positions (up and down). In TTL logic, a "low" voltage will be approximately 0v and a "high" will be approximately +5v (actually anywhere from +2v to +5v).

1.3   Record your observations when a logic probe is used to test the output from the logic switch and it is placed in each of its two positions (up and down). You may need a short piece of jumper wire if the probe tip does not make electrical contact to the switch. If you are unfamiliar with the use of the logic probe, refer to the "Measuring Logic Levels with a Logic Probe" section of this lab.

1.4   Locate the lamp monitors on the digital tester and determine the number of individual lights available. Record your observations of the operation of the lamp monitors by connecting a jumper wire from one of the logic switches to a lamp monitor and then moving the switch between the two logic levels (high and low voltage). In positive logic, a high voltage is referred to as a "1" and a low voltage is referred to as a "0". How does the lamp monitor result compare to the logic probe's?

1.5   Locate the pushbuttons or pulsers on the digital tester and determine the number of individual momentary inputs available. Note the operation of a pushbutton or pulser by connecting it to an unused lamp monitor and then pressing and releasing the button. If your unit has pushbuttons with two complementary outputs available, connect each of the outputs to a separate, unused lamp monitor and then press and release the button.

1.6   Note the operation of the clock output by connecting it to an unused lamp monitor with the clock set at its lowest frequency. If your unit has a clock with two complementary outputs available, connect each of the clock outputs to a separate, unused lamp monitor. Describe the lamp action when the clock frequency is increased.

1.7   Investigate the internal connections of the breadboarding socket shown in the following diagram. Wire the connections to the power supply, to the CLOCK, and to a LOGIC SWITCH as shown in the diagram. Use the logic probe to test the socket holes around the connections that you have made to the socket. Sketch the breadboarding socket and explain its internal electrical conductor pattern. Note: If your breadboarding socket does not have bus strips available as illustrated, connect the power supply instead to other sections of the socket to make your tests.

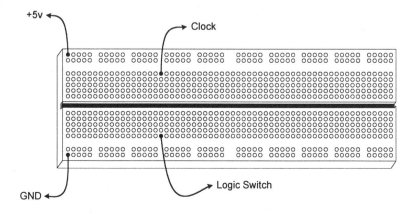

1.8   Construct the logic circuit shown in the following drawing on a breadboarding socket. Two IC chips are used in the circuit. One (the 74LS08) contains AND gates and the other (the 74LS32) contains OR gates. The given circuit will have the same input combination simultaneously applied to an AND gate and an OR gate. The output from each gate will be observed on separate lamp monitors. The output of the AND gate is named X and the output of the OR gate is named Z. Two logic switches (A & B) are used to provide the inputs to the logic circuit. There will be four possible combinations of inputs with the two switches A & B. Be sure to connect power (+5V) and ground to each IC chip. Complete the table below to record the test results. Describe the necessary switch inputs to produce a high output for each of the two gate types (AND & OR).

A	B	X	Z
0	0		
0	1		
1	0		
1	1		

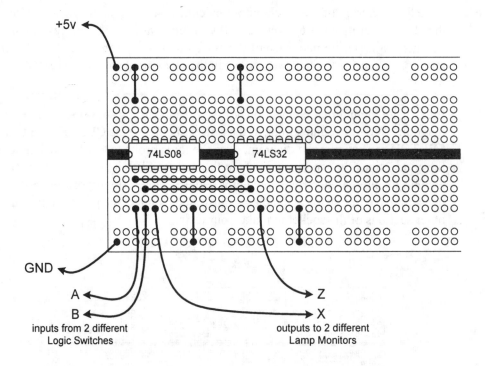

BREADBOARDING COMBINATIONAL LOGIC CIRCUITS

Objectives

- To construct simple combinational logic circuits from a schematic.
- To test simple combinational logic circuits to determine the functional operation of the circuits.
- To identify common logic functions produced by various circuit configurations by the resulting truth table.
- To connect various gates together to create simple logic functions.

Suggested Parts					
74LS00	74LS02	74LS04	74LS08	74LS20	74LS27
74LS32					

Combinational Logic Circuits

Logic Gates

A logic gate is the simplest device used to construct digital circuits. The output voltage or logic level for each type of gate is a function of the applied input(s). Various types of logic gates are available (including inverters, ANDs, ORs, NANDs, and NORs), each with its own unique logic function. Logic circuits are constructed by interconnecting various logic gates together to implement a particular circuit function.

Truth Tables

The logic function for a single gate or a complete circuit using many gates can be easily represented in a logic truth table or a logic expression. The layout for 1 and 2 input variable truth tables is given in Fig. 2-1. Note: L = low logic level voltage and H = high logic level voltage.

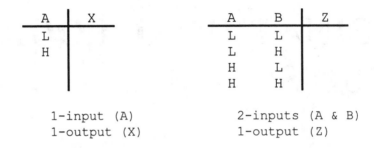

A	X
L	
H	

1-input (A)
1-output (X)

A	B	Z
L	L	
L	H	
H	L	
H	H	

2-inputs (A & B)
1-output (Z)

Fig. 2-1 Truth tables showing possible input combinations

Circuit Breadboarding

When constructing logic circuits, care should be taken to ensure that the parts are not damaged while breadboarding the circuit. Carefully insert the integrated circuit (IC) into the breadboarding socket so that the IC is straddling the center groove on the socket. Make sure that both rows of pins are correctly lined up with the holes in the breadboard, but be careful to avoid bending the IC pins any more than is necessary. The ICs can be safely removed by prying up each end with a screwdriver (or similar tool) or an IC puller. Do not insert or remove ICs with the power applied to the circuit.

A notch or dot at one end of the IC package is used to locate pin 1 of the chip, and the pin numbers then increase in a counterclockwise direction around the device as viewed from the top (see Fig. 2-2). Inserting all ICs with the same orientation for pin 1 will facilitate circuit wiring and troubleshooting.

Determine the pin-out information for <u>each</u> chip by consulting a data book or data sheet for the logic devices used. Notice that power and ground, inputs, and outputs are sometimes located on different pins for different part numbers. Label the pin numbers for each device on a logic circuit schematic to aid in wiring the circuit and in troubleshooting it later if necessary.

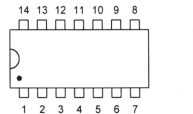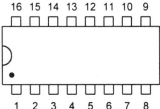

Fig. 2-2 IC pin numbering for 14 & 16 pin DIP packages (top view)

Systematically and carefully wire the circuit with the power off. Wiring errors are the most common source of circuit failure in breadboarding circuits. The jumper wires used to connect the circuit components together should have only about 1/4 inch of the insulation stripped from each end of the wire to avoid inadvertently shorting the wires together. Using short jumper wires will facilitate troubleshooting later if necessary. Double check the wiring against the schematic diagram.

Circuit Testing and Troubleshooting

Verify that the power supply is correctly connected to the circuit before turning it on. Also make sure that the power supply voltage is the proper value (5 volts). Then, and only then, should you turn on the power to the circuit.

```
              Troubleshooting Checklist

  ✓  Do the parts used in the circuit match the
     schematic?
  ✓  Have the pin numbers been identified
     correctly in the schematic?
  ✓  Are the parts inserted correctly in the
     breadboard?
  ✓  Is the correct voltage supply being used to
     power the circuit?
  ✓  Is the power properly connected to each
     chip?
  ✓  Are there any wires shorted to one another?
  ✓  Is the circuit wired correctly?
  ✓  Has the circuit been analyzed correctly?
  ✓  Has the circuit been designed correctly?
```

Breadboarded logic circuits can be manually tested using a digital test system like the one described in Unit 1. Connect a separate logic switch to each of the circuit's inputs and a lamp monitor to the circuit's output. Apply various input logic levels with the switches and monitor the output produced by the circuit with the lamp. List the resultant functional operation of the circuit in a truth table. Determine if the circuit is operating properly.

If the circuit does not function properly or if an IC gets very hot or starts smoking, turn off the power and all signals to the circuit <u>immediately</u>. Use the suggestions given in the Troubleshooting Checklist to troubleshoot the malfunctioning circuit.

<u>Example 2-1</u>

Determine the logic expression for the circuit given in Fig. 2-3 and predict the theoretical operation for that circuit. Also show how to wire this logic circuit on a breadboarding socket.

The schematic contains two 2-input NANDs and one inverter. Therefore, two chips will be needed - the 74LS00, which contains four 2-input NAND gates, and the 74LS04, which contains six inverters. The selected parts will have two extra NANDs and five extra NOTs. The schematic has been annotated with appropriate part numbers and pin numbers. Note that other equally good pin number combinations could have been selected for the implementation of this circuit design. A sketch of the wiring layout for this circuit on a breadboard socket is shown in Fig. 2-4. Note that only a portion of the breadboard socket is shown. The heavy black lines represent the appropriate wiring connections for this schematic.

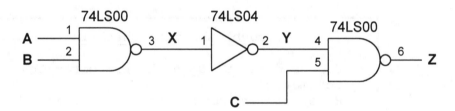

Fig. 2-3 Schematic for example 2-1

The logic expression for each gate output node is determined as follows:

$$X = \overline{A \cdot B}$$

$$Y = \overline{X}$$

$$Z = \overline{Y \cdot C}$$

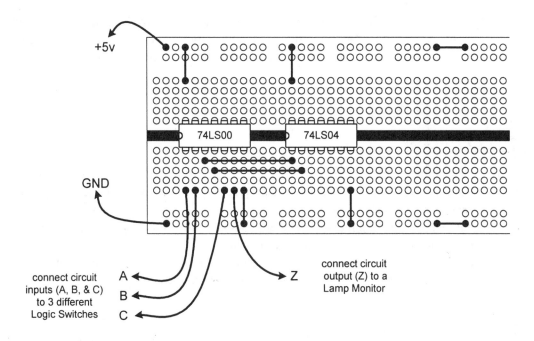

Fig. 2-4 Breadboard wiring of logic circuit in example 2-1

The analysis of the circuit is performed by predicting its theoretical operation in a truth table. The function has three input variables named A, B, and C. Three variables will produce a total of eight input combinations, which are listed in the truth table. Next, each of the gate output nodes are listed and then predicted, as shown in Table 2-1.

A	B	C	X	Y	Z
0	0	0	1	0	1
0	0	1	1	0	1
0	1	0	1	0	1
0	1	1	1	0	1
1	0	0	1	0	1
1	0	1	1	0	1
1	1	0	0	1	1
1	1	1	0	1	0

Table 2-1 Truth table prediction for example 2-1

The overall function of the circuit is represented by output Z for the corresponding inputs A, B, and C. Note that the resultant function for this circuit is equivalent to a 3-input NAND gate and, therefore, the expression would be equivalent to:

$$Z = \overline{A \cdot B \cdot C}$$

Laboratory Projects

Determine the logic expression for each of the following simple logic circuits. Analyze each circuit to predict its theoretical operation. Redraw the schematics for each circuit. Annotate each of the schematics with appropriate component part numbers and pin numbers. Then construct and test each circuit to verify your predictions. Give the theoretical results and test results for each circuit in a logic truth table. Identify any common logic functions produced by the given circuits. Reconcile any differences between the predicted and test results.

2.1   One-chip logic circuits
Construct and test each of the following logic circuits. Each circuit will use only 1 chip.

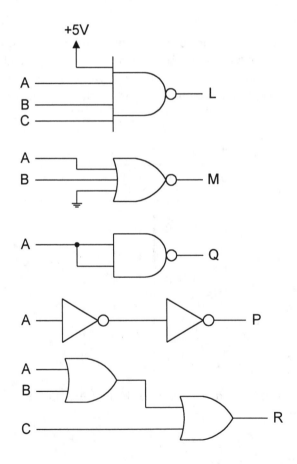

2.2   Two-chip logic circuits
Construct and test each of the following logic circuits. Each circuit will use 2 chips.

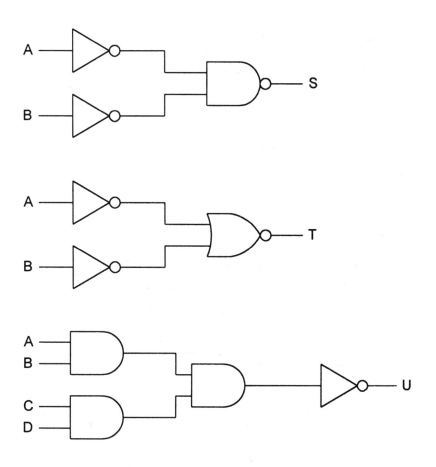

2.3   Three-chip logic circuits
       Construct and test each of the following logic circuits. Each circuit will use 3 chips.

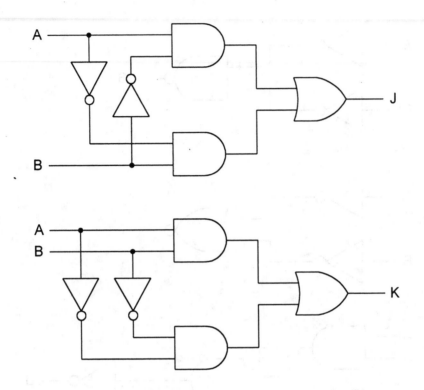

ANALYZING COMBINATIONAL LOGIC CIRCUITS

Objectives

- To analyze combinational logic circuits and predict their operation.
- To construct and test more complex, combinational logic circuits.

Suggested Parts					
74LS00	74LS02	74LS04	74LS08	74LS10	74LS20
74LS32	74LS86				

Combinational Logic Circuit Analysis

The theoretical operation of a combinational logic circuit can be predicted by analyzing the circuit's output for every possible input combination. The circuit analysis for each input combination is performed by determining the resultant output of each gate, working from the input side of the circuit to the output. We will later discover shortcuts to speed up the analysis process.

Logic circuits may be functionally equivalent. They may perform the same function (i.e., their logic truth tables are identical) but be constructed from different logic gates or interconnected in an entirely different manner. In fact we will find that often a complex logic circuit can be replaced with a much simpler one that performs the identical function.

Example 3-1

Analyze the circuit in Fig. 3-1 and determine its truth table.

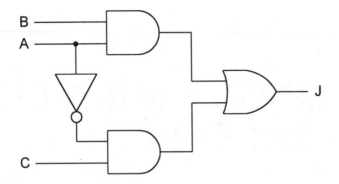

Fig. 3-1 Schematic for circuit analysis example

The logic expression for the given logic circuit is:

$$J \ = \ A \ B \ + \ \overline{A} \ C$$

Since the logic circuit has 3 input variables, a truth table (see Table 3-1) listing all 8 possible combinations is constructed. Create a separate output column in the truth table for each logic gate in the circuit and label the column to represent the gate function or output node name. The resultant output for every gate in the circuit can then be determined for each of the 8 possible input combinations. This gate output information is added to the truth table results for each of the logic gates.

A	B	C	$\overline{A}$	A B	$\overline{A}$ C	J
0	0	0	1	0	0	0
0	0	1	1	0	1	1
0	1	0	1	0	0	0
0	1	1	1	0	1	1
1	0	0	0	0	0	0
1	0	1	0	0	0	0
1	1	0	0	1	0	1
1	1	1	0	1	0	1

Table 3-1 Truth table analysis for example circuit

Laboratory Projects

Combinational circuit analysis and testing
Predict the theoretical operation for each of the following logic circuits. Draw and label circuit schematics with appropriate part numbers (instead of the arbitrary gate names shown) and pin numbers. Then construct and test each circuit. Explain the given titles for each circuit and note any other observations.

3.1   Exclusive-OR circuit

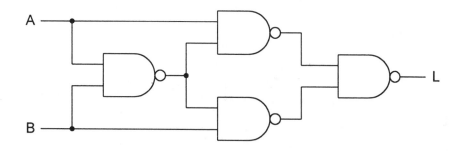

3.2   Exclusive-NOR circuit

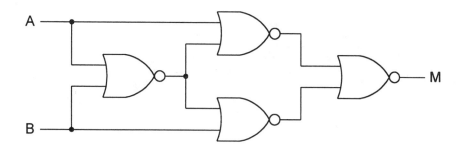

3.3 Inverter circuit

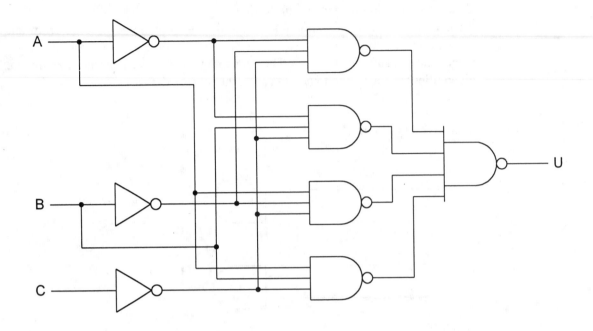

3.4 AND circuit

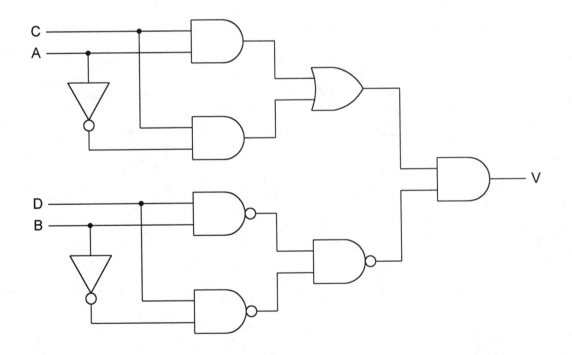

3.5 OR circuit

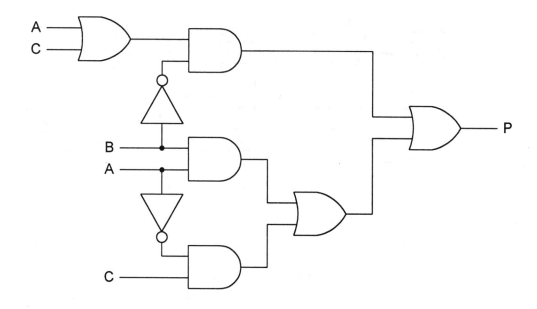

3.6 AND circuit

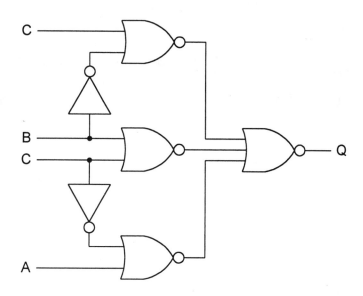

3.7   Multiple-output circuit

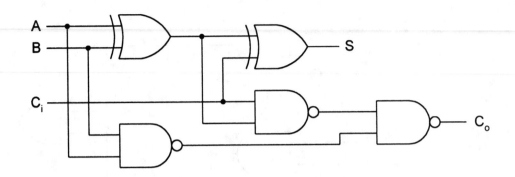

COMBINATIONAL CIRCUIT DESIGN: SOP EXPRESSIONS AND KARNAUGH MAPPING

Objectives

- To write standard sum-of-product (SOP) logic expressions for functions defined in given truth tables.
- To implement SOP logic expressions using standard AND/OR and NAND/NAND circuit configurations.
- To write simplified SOP logic expressions using Karnaugh mapping techniques.
- To design a combinational logic circuit that will perform a stated task by first defining the logic function with a truth table and then determining the simplified circuit solution using Karnaugh mapping.

Suggested Parts					
74LS00 74LS393	74LS04	74LS08	74LS10	74LS20	74LS32

Combinational Circuit Design

SOP Expressions

The most commonly used format for writing a logic expression is a standard form called sum-of-product (SOP). SOP expressions can be quickly written from a truth table and they are easily implemented using a two-level (not counting inverters that may be needed) gate network. Conversely, these standard circuits that are used to implement SOP functions can be quickly analyzed just by inspection. A sum-of-product expression consists of two or more product (AND) terms that are ORed together. The SOP expression is obtained from a truth table by writing down all of the product terms (also called minterms) whose outputs are high for the desired function and then ORing them together. The resultant SOP expression can be directly implemented with either AND/OR or NAND/NAND circuit designs.

Karnaugh Mapping

The basic procedure for combinational logic circuit design is to develop first the truth table that defines the desired function and then from the table, write a simplified SOP expression. The expression can be simplified using various techniques (such as Boolean algebra, Karnaugh mapping, etc.). Karnaugh mapping is a simple and fast procedure for reducing SOP logic expressions and thereby also reducing the implemented circuit's complexity and cost. In Karnaugh mapping, the function is defined graphically. The relationships between the function's inputs and the output are plotted in a Karnaugh map (K-map). This will be the same information that would be listed in the truth table for the function. The input variables must be labeled on the K-map in a very systematic fashion. If the K-map is not properly labeled, the function cannot be correctly simplified and the resulting design will be wrong. With K-mapping, the function reduction is accomplished by forming appropriate groupings of 1s in the output. Then identify the common input variables for the group and write the indicated product term. Karnaugh mapping can best be applied to functions with 5 or fewer input variables.

Example 4-1

Design a combinational circuit that will indicate the majority result of 3 individuals voting.

First, define the problem in a truth table as shown in Table 4-1.

A	B	C	V
0	0	0	0
0	0	1	0
0	1	0	0
0	1	1	1
1	0	0	0
1	0	1	1
1	1	0	1
1	1	1	1

Table 4-1 Truth table for example 4-1

The unsimplified SOP expression for example 4-1 would be:

$$V = \overline{A} B C + A \overline{B} C + A B \overline{C} + A B C$$

The Karnaugh map for this function is plotted in Fig. 4-1.

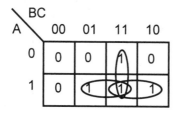

Fig. 4-1 Karnaugh map for example 4-1

The simplified SOP expression for this function would be:

$$V = B C + A C + A B$$

The simplified SOP expression can be easily implemented with a NAND/NAND circuit arrangement using 2 chips (7400 and 7410). The circuit schematic for this solution is given in Fig. 4-2.

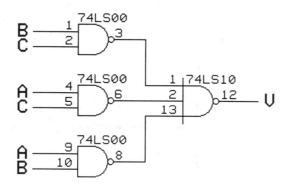

Fig. 4-2 Schematic for simplified SOP solution to example 4-1

Example 4-2

Design a simplified logic circuit to implement the function W defined in the truth table given in Table 4-2.

A	B	C	D	W
0	0	0	0	0
0	0	0	1	1
0	0	1	0	1
0	0	1	1	1
0	1	0	0	0
0	1	0	1	0
0	1	1	0	0
0	1	1	1	0
1	0	0	0	1
1	0	0	1	1
1	0	1	0	0
1	0	1	1	0
1	1	0	0	0
1	1	0	1	0
1	1	1	0	0
1	1	1	1	0

Table 4-2 Truth table for example 4-2

The output produced for the function W is plotted in a Karnaugh map in Fig. 4-3. Then appropriate groups of 1s are identified in the K-map to create the SOP expression for W. The two simplified expressions given in Fig. 4-3 can be obtained with K-mapping. The groupings of 1s shown in the K-map are represented by the first equation. The two solutions each require the same number of gates or chips and so either simplified SOP expression can be implemented. See Fig. 4-4a and b for the two solution schematics.

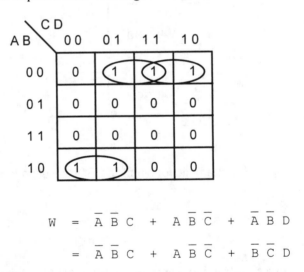

$$W = \overline{A}\,\overline{B}\,C + A\,\overline{B}\,\overline{C} + \overline{A}\,\overline{B}\,D$$

$$= \overline{A}\,\overline{B}\,C + A\,\overline{B}\,\overline{C} + \overline{B}\,\overline{C}\,D$$

Fig. 4-3 Karnaugh map and simplified expressions for example 4-2

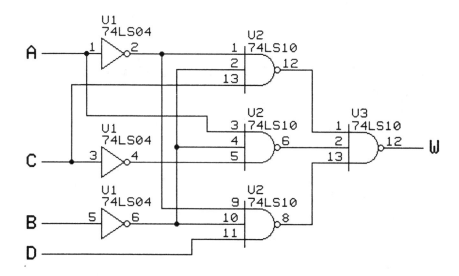

Fig. 4-4a Schematics for simplified SOP solutions to example 4-2

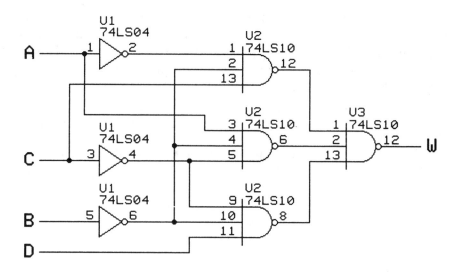

Fig. 4-4b Schematics for simplified SOP solutions to example 4-2

Laboratory Projects

Design logic circuits to perform each of the following functions. Define the problem with a truth table and then use K-mapping to write the simplified SOP expression. Draw and label the schematics to implement your simplified designs. Construct, test, and verify the operation of your design.

4.1   Two-input multiplexer
Design a circuit whose output (Y) is equivalent to one of two possible data inputs (A or B). A control input (S) selects either the data on the A input (if S is low) or the data on the B input (if S is high) to be routed to the single output line.

4.2   Three-bit equality detector
Design a 3-bit equality detector circuit that will output a low whenever the 3 input bits are <u>all</u> at the same logic level.

4.3   Elevator control
Design an elevator control system for a large building that has 5 elevators. Four of the elevators are turned on all of the time, while the fifth is activated only if a majority of the other 4 are being used (to save energy). The control system will have an input for each of the 4 primary elevators to indicate that that elevator is being used (with a logic "1"). A high output from the control system will activate the fifth elevator for its use.

4.4   Greater than 9 detector
Design a circuit whose output will be high if the 4-bit data input is a value greater than 9.

4.5   Window detector
Design a circuit whose output will be low for all 4-bit input combinations that meet the following criteria:
$$4 \; < \; I \; < \; 11$$
where I represents the 4-bit input value.

4.6   Two-bit comparator
Design a comparator circuit to compare two 2-bit numbers (A1 A0 and B1 B0). The circuit will have two output signals: GE and LT. GE will be high to indicate that the 2-bit A value is equal to or greater than the 2-bit B value. LT will be high if A < B.

4.7   Alarm circuit

Design an alarm circuit to be used in a process control system. Temperature (T), pressure (P), flow (F), and level (L) of a fluid are each monitored by separate sensor circuits that produce a <u>high</u> logic output signal when the following indicated <u>physical</u> conditions exist:

```
high fluid temperature
high fluid pressure
low fluid flow rate
low fluid level
```

The alarm circuit output (A) should be <u>high</u> if any of the following <u>physical</u> conditions exist in the system:

```
(1)   the pressure is high when the flow is low
(2)   the temperature is high when either the pressure
      is high or the level is low
```

Be sure to identify the correct physical conditions for the alarm in the logic truth table.

4.8   Prime number detector

Design a 4-bit prime number detector circuit. The 4-bit input will allow the binary numbers for 0 through 15 to be applied to the circuit. The output should be high only if prime numbers (1, 2, 3, 5, 7, 11, 13) are being input to the detector circuit.

4.9   Multiplier circuit

Design a multiplier circuit that will output the product of any 3-bit input number (0 through 7) multiplied by the constant 3. Hint: This circuit will have several output bits and each of the outputs will have to be mapped <u>separately</u>. Each output bit represents a circuit that must be constructed.

4.10   Digital switcher

Design a digital signal switching circuit that has two outputs (X and Y). For each of the two outputs, the circuit can select from two different signal sources (inputs A or B). The two input signals will be obtained from a binary counter chip, the 74LS393, as shown in the diagram below. The 74LS393 is being used as a 2-bit counter in this application. Use a low frequency such as 1 Hz for the CLOCK input to the counter. Do not forget to ground the CLR control on the 74LS393 or the counter will stay at zero and not count. The input signal selection is controlled by the signals C and D. The two control signals will be obtained from two logic switches. The circuit function is described in the following truth table. Test your design using lights to monitor X and Y. If a dual-trace oscilloscope is available, increase the CLOCK frequency to approximately 1 KHz and monitor the two output signals (X and Y) on the scope. Hint: Expand the function table to show the 16 combinations that are possible with the four inputs D, C, B, and A and K-map for each of the functions X and Y.

D	C	X	Y
0	0	A	B
0	1	A	A
1	0	B	B
1	1	B	A

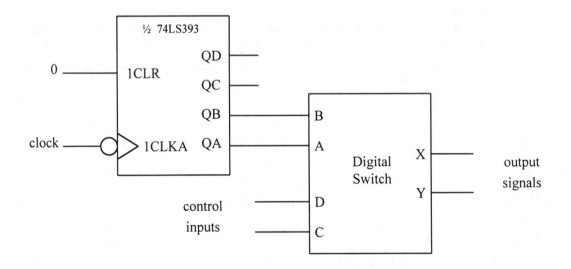

UNIT 5

CIRCUIT MINIMIZATION WITH BOOLEAN ALGEBRA

Objectives

- To simplify logic circuits using Boolean algebra techniques.
- To determine the minimum number of logic chips needed to implement a logic function.

Suggested Parts				
74LS00	74LS02	74LS04	74LS10	74S260

Logic Circuit Minimization

The various laws and theorems of Boolean algebra can be systematically applied to logic expressions in order to manipulate and/or simplify the expressions. Of course, extreme care must be exercised to correctly apply the laws and theorems or the resultant expression will not be equivalent to the original function. There are often many alternative Boolean algebra techniques that can be applied in the process of logic circuit simplification.

While a given logic expression is manipulated into different equivalent expression forms, the corresponding circuit implementations of each expression can be analyzed to determine the quantity of chips necessary for the circuit construction. The design that uses the minimum number of available chips may be selected as an optimum design choice (based on the amount of board space required by the circuit, the cost of the needed parts, etc.). This process generally requires some trial and error in order to determine the minimum chip solution.

Example 5-1

Redesign the circuit in Fig. 5-1 so that its function may be implemented with a minimum number of chips.

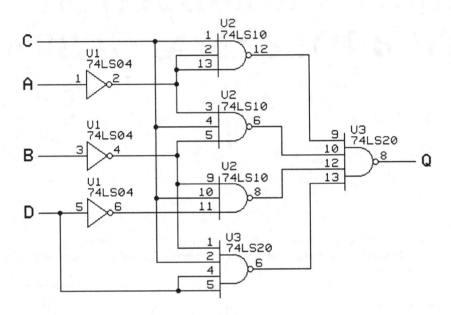

Fig. 5-1 Logic circuit for example 5-1

First, determine the original circuit's logic expression and then, using Boolean algebra techniques, reduce the expression to an equivalent simplified form. Determine the necessary chips for implementation of the simplified expression. Continue manipulating the expression and selecting appropriate chips for implementation until a minimal design is found.

$$Q = \overline{A}\,C \quad \overline{A}\,\overline{B}\,C \quad \overline{B}\,C\,\overline{D} \quad \overline{B}\,C\,D\,D$$
original circuit expression
(uses 3 chips: 7404, 7410, 7420)

$$Q = \overline{A} \, \overline{C} + \overline{A} \, \overline{B} \, \overline{C} + \overline{B} \, \overline{C} \, \overline{D} + \overline{B} \, \overline{C} \, \overline{D}$$

$$Q = \overline{A} \, C + \overline{A} \, \overline{B} \, C + \overline{B} \, C \, \overline{D} + \overline{B} \, C \, D$$

$$Q = \overline{A} \, C \, (\, 1 + \overline{B} \,) + \overline{B} \, C \, (\, \overline{D} + D \,)$$

$$Q = \overline{A} \, C + \overline{B} \, C \quad \Leftarrow \text{ possible solution?}$$
$$\text{(uses 3 chips: 7404, 7408, 7432)}$$

$$\overline{\overline{Q = \overline{A} \, C + \overline{B} \, C}}$$

$$Q = \overline{A} \, C + \overline{B} \, C$$
$$\text{try another alternative route}$$

$$Q = \overline{\overline{A} \, C} \quad \overline{\overline{B} \, C} \quad \Leftarrow \text{ possible solution?}$$
$$\text{(uses 2 chips: 7400 \& 7404)}$$

$$Q = \overline{A} \, C + \overline{B} \, C$$
$$\text{try another route from here}$$

$$Q = C \, (\, \overline{A} + \overline{B} \,) \quad \Leftarrow \text{ possible solution?}$$
$$\text{(uses 3 chips: 7404, 7408, 7432)}$$

$$Q = C \, \overline{\overline{A} \, \overline{B}} \quad \Leftarrow \text{ possible solution?}$$
$$\text{(uses 2 chips: 7404 \& 7408)}$$

$$Q = C \, \overline{\overline{A} \, \overline{B}} \quad \Leftarrow \text{ possible solution?}$$
$$\text{(uses 1 chip: 7400)} \quad \therefore \; \textbf{best solution!}$$

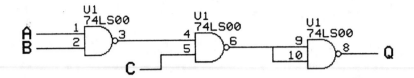

Fig. 5-2 Minimized circuit solution

Laboratory Projects

Redesign each of the following logic circuits so that a minimum number of available chips will be used to implement the logic functions of each circuit. Compare the number of chips required for the original circuits with your new equivalent designs. Construct and test each of your simplified designs.

5.1   Circuit 1

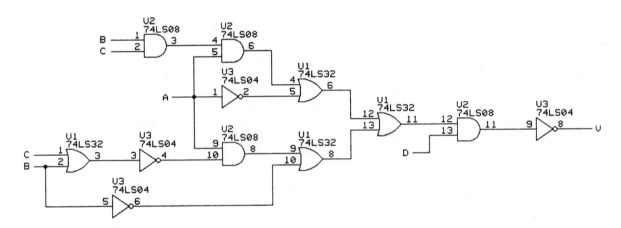

5.2   Circuit 2

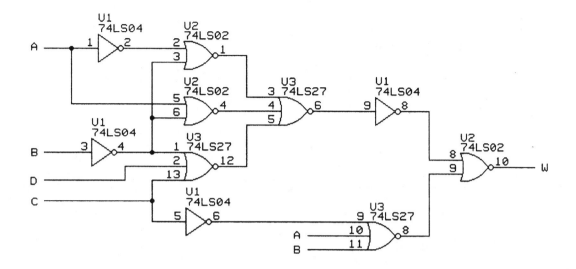

5.3   Circuit 3

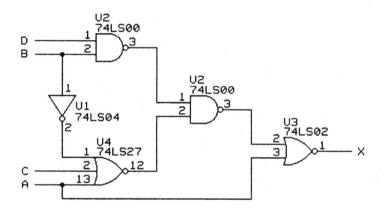

5.4 Circuit 4

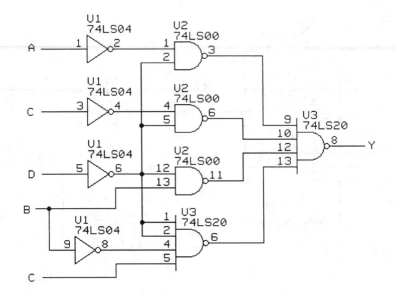

5.5 Circuit 5

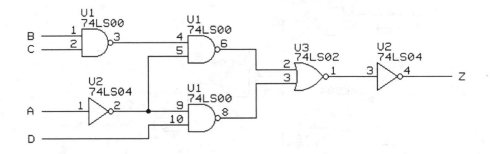

UNIT 6

COMBINATIONAL CIRCUIT DESIGN USING PROGRAMMABLE LOGIC DEVICES

Objectives

- To implement combinational logic circuits using programmable logic devices (PLDs).
- To create PLD logic description files that define combinational circuit designs using Boolean expressions and truth tables.
- To use logic compiler software to create JEDEC files for programming PLDs.

Suggested Part
GAL16V8

Programmable Logic Devices

With programmable logic devices (PLDs), logic circuit designers can go from a conceptual design to customized functional parts in a matter of minutes. A PLD is a digital IC that is capable of being programmed to provide a specific logical function. The PLD family of devices consists of a variety of device architectures and configurations. The most common PLDs are based on the familiar AND/OR logic gate array in which the specific inputs to each AND gate are programmed to achieve the desired function. Sets of available AND gates in the PLD are internally connected to different OR gates to produce the needed outputs. The programmable AND/OR gate configuration is used to implement sum-of-product functions and, since the SOP form can be used to express any Boolean function, PLD designs are limited only by the number of terms available in the arrays. Many different PLD part numbers are available that provide a wide variety of choices in the number of inputs and outputs that are available, the number of product terms that can be handled, the output assertion level, the ability for outputs to be tristated, and the ability to produce registered outputs. The programmable flexibility of PLD devices typically allows circuit designers to replace many different standard SSI/MSI chips with a single PLD package. PLD devices are available for either one-time-only programming (commonly referred to as PALs) or are erasable and reprogrammable (EPLDs and EEPLDs).

The Lattice Semiconductor GAL16V8 is an example of an electrically erasable PLD. The GAL16V8 is quite versatile. This device can be programmed for a maximum of 8 outputs or a maximum of 16 inputs (although not that many of each simultaneously). Each of the outputs can accommodate up to 8 product terms. The GAL (generic array logic) chip has a flexible output structure called an output logic macrocell. The output logic macrocell (OLMC) allows a number of options to be programmed into the device. Each of the outputs can be programmed to be active high or active low, to be combinational or registered, and to be tristated or normal.

The GAL16V8 is a CMOS device and, as such, should be handled carefully. CMOS devices can be easily damaged by static electricity. Some recommended precautions for handling CMOS devices include storing the chip in conductive foam when not in a circuit, wearing an anti-static wrist strap, turning off the power when inserting or removing the chip, and connecting all unused IC pins to either ground or Vcc when the chip is inserted in a circuit.

Implementing a logic circuit with a PLD consists of developing the appropriate design information (Boolean equations, truth tables, or state machine description) that is then entered into a (computer) logic description file using a text editor (or a word processor program). Input and output pin assignments and the selected target device for the design are also included in the logic description file. A logic compiler running on a computer is then used to transform the circuit description file into a standard output form called a JEDEC file. The JEDEC file is then downloaded to a PLD programmer that is used to program the design into the PLD. See the summary procedure in Fig. 6-1. The PLD is then ready to be used in your application. Design changes can be implemented simply by altering the logic description file with the text editor, re-compiling the logic description file, and then reprogramming the same PLD (if it is electrically erasable like the GALs). The old EEPLD configuration is erased automatically by the programmer hardware when it is reprogrammed.

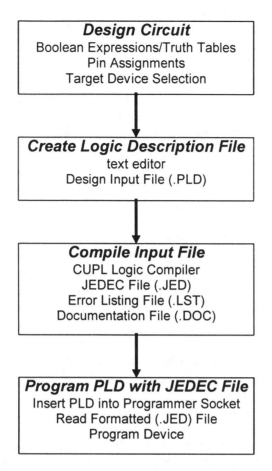

Fig. 6-1 Summary PLD design procedure

Logic Compiler Operation

An input file that contains the logic description for a PLD design is first created using a text editor. A logic compiler is then used to translate the logic description file information into a standard output file format called a JEDEC file (filename.JED) that, in turn, is used to program the PLD device for the desired function. Several logic compilers are available to perform this translation task. They are very similar in their operation and features. One example, CUPL (from Logical Devices, Inc.), is a very powerful but easy to learn logic compiler. The Windows version of this compiler (CUPLWIN) provides a standard Windows-type user interface to operate the software. All design file creation, editing, and compilation is handled through CUPLWIN. Design simulation can also be performed with CUPLWIN (see Appendix B). An educational version of CUPLWIN is available through Prentice-Hall, the publisher of this lab manual. The educational version of the compiler has the same functionality as the commercial version of the software but has a limited library of parts that may be used as target devices. A DOS version of the software with a menu-driven user-interface is also available from Logical Devices. Downloading of the compiler's

JEDEC output file and device programming is handled separately and the procedure is dependent upon the specific programmer used.

The compiler's Main menu choices are: File, Edit, Option, Run, Utilities, Project, Windows, and Help. Only the principal menu choices and commands will be mentioned in this manual. Each menu can be pulled-down by clicking the desired choice with the left mouse button. The File menu lists the familiar Windows' commands: New, Open, Save, Save As, Print, and Exit. New will display a template for a logic description file (see the next section for example files) that may be modified and then saved as a different file name. Open will display a dialog box listing the .PLD files in the WinCUPL directory (or the Working Directory listed by the user under Preferences in the Option menu). Other file types (extensions) can be listed in the Open a File dialog box by clicking on the List Files of Type arrow and choosing the desired file type filter. The Edit menu includes the commands: Cut, Paste, Copy, and Delete. The Windows menu includes: Cascade, Tile, and a list of the files that have been opened. The Help menu can be consulted for additional error message (by message number) information. Note that many of the commands have short cut keys that may be used.

After the logic description file has been created, but before it can be compiled by CUPL, appropriate compiler options must be selected. Many choices are optional, but the general procedure is:

- From the Main menu, select Option and then choose Compiler Options. The Compile dialog box will open. Click the Output file button.
- Select "list" to create an output file (filename.LST) that is saved on the disk. This error listing file contains a copy of the input file annotated with any error messages and markers produced by the compiler due to syntax errors encountered in the input file. The error messages will also be scrolled in the Message window, but it is often easier to debug the input file by looking at the "LST" file. The LST file can be opened with "File" - "Open" and using the List Files of Type filter in the Open a File dialog box.
- Select "equations" under Doc File Options. The documentation file (filename.DOC) contains the SOP equations generated by the compiler and a symbol table of all variables used in the input file.
- Select JEDEC as the Download Output Format. Note that if "Format j;" is specified in the header section of the logic description file (see examples in the next section) it is not necessary to specify JEDEC again in the Output Format dialog box. You may find it more convenient to modify the template file by adding "Format j;" (lower case j) to the header section.
- Click OK and return to the Compile dialog box.
- Select the desired Logic Minimization algorithm. CUPL provides five different choices of minimization algorithms for the user to select from including: None, Quick, Quine-McCluskey, Presto, and Expresso. The most thorough but slowest algorithm is Quine-McCluskey. The fastest minimization algorithm is Quick.
- The Select Files button can be used to specify the input file to be compiled. However, if the desired file is in the active window (more than one input file may be opened at a time), it will automatically be selected as the input file for the compiler. **It is important to remember that the file is compiled from the disk**

and not from the screen. If you make any edits of the file, be sure to save them before compiling!

- The Select Device button <u>can</u> be used to specify the target device. However, the device can also be specified with "Device G16V8A;" in the input file header (see the examples in the next section). Again you may find it more convenient to modify the template file by adding "Device G16V8A;" to the header section.
- Click OK and return to the main CUPL screen.
- After saving any edits to the input file, pull-down Run from the Main menu and select Device Specific Compile. The Compile Status window will open and the compiler will start. Compiler progress will be displayed in the Status window and in the Message window. When finished, the top line in the Status window will indicate either "Compilation successful!" or "Errors encountered." Click OK. If successful, you should have the output file (filename.JED) stored on the disk. If syntax errors were detected by the compiler, the error messages can be read in the Message window(by scrolling it) or in the filename.LST file. Correct any errors in the input file, save the corrections, and re-compile. When you have a successful compilation, download the JEDEC file to the programmer and program a GAL chip.

CUPL Logic Description File

The CUPL logic description file contains the data that describes the desired PLD logic design. CUPL assumes that the logic description file is named with the extension ".PLD". A text editor, such as the one included in CUPLWIN, is used to create this logic description file. The essential logic description file information includes:

```
HEADER INFORMATION
PIN ASSIGNMENTS
LOGIC DESCRIPTION
```

The header information section is normally placed at the beginning of the logic description file. Each statement in the header section will begin with a keyword, followed by any valid ASCII characters, and will end with a semicolon. The header consists of the keywords and appropriate data shown in Fig. 6-2.

The pin assignment statements are used to assign input and output variable names to desired pin numbers on the PLD. These assignments generally have a great deal of flexibility but must be made in accordance with the architectural capabilities of the target PLD device.

The logic description section of the PLD logic description file provides the functional definition of the logic circuit design. The circuit design can be described with a combination of three formats: state machine description (for sequential circuits), truth table, or Boolean equation.

In addition, other types of information will typically be included to make the design file easier to interpret and to develop. This includes:

```
COMMENTS
DECLARATIONS AND INTERMEDIATE VARIABLE DEFINITIONS
```

The PLD design logic description file should be annotated with comments for clarity. The logic of the design will be much more understandable with appropriate comments. Comments will be ignored by the logic compiler.

```
NAME        logic description filename [8 characters max]
PARTNO      part number for design [not PLD part number]
REVISION    file revision number used to track design updates
            [start with 01 and increment each time file is
            altered]
DATE        date of last file changes
DESIGNER    designer's name
COMPANY     company's name [for documentation]
ASSEMBLY    PC board name where PLD will be used
LOCATION    schematic identifier for PLD location
DEVICE      default device type [can be overridden to select
            alternate device during compilation of logic
            description file]
FORMAT      instructs compiler to produce correct format output
            file automatically for programming PLD
            [use j (lower case) to produce JEDEC output file]
```

Fig. 6-2 CUPL header information

Bits may be grouped together and intermediate variables can be defined to make writing the logic description for a design much easier and clearer. This information is placed in the declarations and intermediate variable definitions section of the design logic description file.

The fundamental language elements for CUPL logic description files are summarized in Fig. 6-3a and b.

CUPL LANGUAGE ELEMENTS

Variables:
- Specify device pins, internal nodes, constants, input signals, output signals, intermediate signals.
- Start with a numeric digit, alphabet character, or underscore.
- Must include at least one alphabet character.
- Are case sensitive.
- Cannot have spaces within the variable name.
- May be up to 31 characters long.
- Cannot contain any reserved symbols.
- Cannot be the same as a reserved keyword.
- Can be indexed variables -- variable names that end in a decimal number between 0 and 31.

Reserved words:

APPEND	ELSE	JUMP	PRESENT
ASSEMBLY	FIELD	MACRO	REV
ASSY	FLD	MIN	REVISION
COMPANY	FORMAT	NAME	SEQUENCE
CONDITION	FUNCTION	NODE	SEQUENCED
DATE	FUSE	OUT	SEQUENCEJK
DEFAULT	IF	PARTNO	SEQUENCERS
DESIGNER	LOC	PIN	SEQUENCET
DEVICE	LOCATION	PINNODE	TABLE

Reserved symbols:

&	#	(	)	–	@	*
+	[	]	/	^	:	.
/*	*/	=	!	;	,	..
$	'					

Numbers:

Values may be from 0 to $2^{32}-1$.

Default base for numbers is hexadecimal except for device pin numbers and indexed variables, which are decimal. Binary, octal, and hexadecimal can have don't care values (X) and numerical values.

Base	Prefix (use upper- or lower-case)
binary	'b'
octal	'o'
decimal	'd'
hexadecimal	'h'

Fig. 6-3a Summary of CUPL language elements

CUPL LANGUAGE ELEMENTS

Logical operators
     &   =   Logical AND   #   =   Logical OR
     $ =   Logical XOR   !   =   Logical NEGATION
Comment indicators
     /* =   Start comment   */ =   End comment

Intermediate variables
     You can arbitrarily create and define a symbolic name as
     follows:
               MEMREQ = MEMW # MEMR;
     where MEMREQ does not appear as a pin variable name. MEMREQ
     can then be used in expressions for other variables. The
     value
     "MEMW # MEMR" will be substituted wherever MEMREQ is used.

List notations
     You can represent groups of variables in a shorthand list
     notation by using the following formats:
               [var1, var2, ..., varN] as in [MEMR,MEMW,IOR,IOW]
                              or
               [varN..0] as in [A7..0], which is equivalent to
               [A7,A6,A5,A4,A3,A2,A1,A0]

Bit fields
     A bit field is a declaration of a group of bits that is
     represented by a single symbolic name. Bit fields are
     declared as follows:
               FIELD IOADR = [A7..0];
     where IOADR can then be used in expressions instead of
     [A7..0].

Equality operator
     The equality operator symbol is a colon ":". This operator
     compares bit equality between a set of variables (or a bit
     field) and a constant value or a list of constant values.
     For example, if IOADR represents a set of eight bits or
     variables, then:
               IOADR:C3
     would be true if the bit field named IOADR were equal to the
     hex value C3. Also:
               IOADR:[10..3F]
     would be true if IOADR were in the range of hex values 10
     through 3F.

Fig. 6-3b Summary of CUPL language elements

Example 6-1

Design a logic circuit that will detect various input conditions for a 4-bit input value (DCBA). The output GT9 will be high if the input value is greater than 9. Another output, LT4, will be low if the input value is less than 4. The output signal RNG will be high if the input is greater than 6 and less than 11. And the output signal TEN will be low if the input is equal to 10.

The first step is to define the problem in a truth table as shown in Table 6-1 and then determine appropriate signal assertion levels and logic expressions. Signals may be asserted either as active-high or active-low signals. To develop the PLD design using a high level compiler such as CUPL, the active level of the signals are determined separately from the logic expression that determines the signal activation. The active level of the signal is independent (and can be easily changed) of the expression that describes activation. **The logic expression determines when the function is to be asserted while the active level given in the pin assignment determines the voltage level of the signal when it is asserted.**

D	C	B	A	GT9	LT4	RNG	TEN
0	0	0	0	0	0	0	1
0	0	0	1	0	0	0	1
0	0	1	0	0	0	0	1
0	0	1	1	0	0	0	1
0	1	0	0	0	1	0	1
0	1	0	1	0	1	0	1
0	1	1	0	0	1	0	1
0	1	1	1	0	1	1	1
1	0	0	0	0	1	1	1
1	0	0	1	0	1	1	1
1	0	1	0	1	1	1	0
1	0	1	1	1	1	0	1
1	1	0	0	1	1	0	1
1	1	0	1	1	1	0	1
1	1	1	0	1	1	0	1
1	1	1	1	1	1	0	1

Function	Active-level for output	Logic equation to produce active output
GT9	high	$D\,B + D\,C$
LT4	low	$\overline{D}\ \overline{C}$
RNG	high	$D\ \overline{C}\ \overline{B} + D\ \overline{C}\ \overline{A} + \overline{D}\ C\ B\ A$
TEN	low	$D\ \overline{C}\ B\ \overline{A}$

Table 6-1 Truth table, signal assertion levels, and equations for PLD example 6-1

The CUPL logic description file shown in Fig. 6-4 has been created for this PLD design. The logic description file has been named LAB-EX1.PLD. Appropriate design information has been provided in the header section, which is given first in the file. Note that a specific target device has been selected for the design using the header entry "Device G16V8A;" (G16V8A is CUPL's abbreviation for a GAL16V8A). The header entry "Format j;" directs the logic compiler to produce a JEDEC output file. These options then do not need to also be specified under Compiler Options. Comments have been included in the logic description file to help document the design. Applicable input and output pins have been selected and declared in the logic description file. Note that the active output level is also indicated by the presence or absence of the negation symbol "!" in the output pin assignment statements. The logic equation section provides the necessary logic expression that describes when each of the outputs is to be asserted. Every statement line ends with a semicolon ";". Comment lines do not end with semicolons since they are ignored by the compiler.

```
Name        LAB-EX1;
Partno      L105-1;
Date        06/30/93;
Revision    01;
Designer    Greg Moss;
Company     Digi-Lab, Inc.;
Assembly    Example Board;
Location    U101;
Device      G16V8A;
Format      j;

        /*****************************************/
        /* example circuit design using a PLD    */
        /* circuit detects various input values  */
        /*    and produces multiple outputs      */
        /*****************************************/

            /**   Input pins   **/
Pin  1  =  D;
Pin  2  =  C;
Pin  3  =  B;
Pin  4  =  A;

            /**   Output pins   **/
Pin 12 =  GT9;   /* GT9 hi output if DCBA > 9     */
Pin 13 = !LT4;   /* LT4 low output if DCBA < 4    */
Pin 14 =  RNG;   /* RNG hi out if 6 < DCBA < 11   */
Pin 15 = !TEN;   /* TEN low output if DCBA = 10   */

            /**   Logic Equations   **/
GT9  =  D & B  #  D & C;
LT4  =  !D & !C;
RNG  =  D & !C & (!B # !A)  #  !D & C & B & A;
TEN  =  D & !C & B & !A;
```

Fig. 6-4 CUPL logic description file for example 6-1

```
**************************************************************
                        LAB-EX1
**************************************************************

CUPL(WM)          4.7a  Serial# MW-66999998
Device            g16v8as  Library DLIB-h-36-2
Created           Mon Dec 30 13:53:39 1996
Name              LAB-EX1
Partno            P105-1
Revision          01
Date              06/30/93
Designer          G. Moss
Company           Digi-Lab, Inc
Assembly          Example Board
Location          U101


===============================================================
                   Expanded Product Terms
===============================================================

GT9 =>
    B & D
  # C & D

LT4 =>
    !C & !D

RNG =>
    !A & !C & D
  # !B & !C & D
  # A & B & C & !D

TEN =>
    !A & B & !C & D


===============================================================
                       Symbol Table
===============================================================

Pin Variable                                Pterms   Max     Min
Pol   Name         Ext     Pin     Type     Used    Pterms  Level
---  --------      ---     ---     ----     ------  ------  -----
      A                     4       V         -       -       -
      B                     3       V         -       -       -
      C                     2       V         -       -       -
      D                     1       V         -       -       -
      GT9                   12      V         2       8       1
  !   LT4                   13      V         1       8       1
      RNG                   14      V         3       8       1
  !   TEN                   15      V         1       8       1

LEGEND  D : default variable      F : field    G : group
        I : intermediate variable N : node     M : extended node
        U : undefined             V : variable  X : extended variable
        T : function
```

Fig. 6-5 CUPL ".DOC" output file for example 6-1

The documentation output file shown in Fig. 6-5 was produced by CUPL (with "equations" selected under Doc File Options in Compiler Options - Output Format) and saved on the disk. The JEDEC output file for a GAL16V8 was created by the compiler. The ".JED" file is then downloaded to a PLD programmer to program a GAL16V8, which has been inserted in the programmer's socket.

Another possible solution to Example 6-1 is shown in Fig. 6-6. In this logic description file, the set of input variables (D, C, B, A) is assigned the name INPUTS (an arbitrary, but appropriate sounding name) in the field statement. INPUTS now can be used to represent the four input bits. The logic equations then can easily be shortened using the equality operator. Whenever INPUTS is equal to the specified constant or range of constants, the resultant expression will be true.

```
Name        LAB-EX1A;
Partno      L105-1;
Date        06/30/93;
Revision    01;
Designer    Greg Moss;
Company     Digi-Lab, Inc.;
Assembly    Example Board;
Location    U101;
Device      G16V8A;
Format      j;

        /*******************************************/
        /* example circuit design using a PLD    */
        /* circuit detects various input values  */
        /*   and produces multiple outputs       */
        /*******************************************/

                /**  Input pins  **/
Pin  1  =  D;
Pin  2  =  C;
Pin  3  =  B;
Pin  4  =  A;

                /**  Output pins  **/
Pin 12 =  GT9;   /* GT9 hi output if DCBA > 9     */
Pin 13 = !LT4;   /* LT4 low output if DCBA < 4    */
Pin 14 =  RNG;   /* RNG hi out if 6 < DCBA < 11   */
Pin 15 = !TEN;   /* TEN low output if DCBA = 10   */

            /**    Declarations and Intermediate
                   Variable Definitions        **/
field INPUTS   =   [D,C,B,A];  /* assigns the set of
                       bits to the name INPUTS */

            /**  Logic Equations  **/
GT9   =   INPUTS:[A..F];       /* default is hex numbers   */
LT4   =   INPUTS:[0,1,2,3];
RNG   =   INPUTS:[7..A];
TEN   =   INPUTS:'d'10;        /* 'd' specifies as decimal */
```

Fig. 6-6 Alternate CUPL logic description file for example 6-1

Example 6-2

Design a logic circuit that will output a binary value equal to the square of a 4-bit input value.

```
Name       SQUARE;
Partno     L105-2;
Date       02/04/91;
Revision   03;
Designer   Greg Moss;
Company    Digi-Lab, Inc.;
Assembly   Generator board;
Location   U210;
Device     G16V8A;
Format     j;

/**************************************************/
/* Generates the square of an input value using */
/* the truth table design entry technique.      */
/**************************************************/

/**  Inputs  **/
Pin [2..5] = [I3..0];     /* Data Input Value    */

/**  Outputs  **/
Pin [12..19] = [N0..7];   /* Square Output Value */

/** Declarations and Intermediate
              Variable Definitions **/
field data_in = [I3..0];
field square = [N7..0];

/** Logic Description  --  truth table format **/
table   data_in  =>   square   {
                 0   =>   'd'0;
                 1   =>   'd'1;
                 2   =>   'd'4;
                 3   =>   'd'9;
                 4   =>   'd'16;
                 5   =>   'd'25;
                 6   =>   'd'36;
                 7   =>   'd'49;
                 8   =>   'd'64;
                 9   =>   'd'81;
                 A   =>   'd'100;
                 B   =>   'd'121;
                 C   =>   'd'144;
                 D   =>   'd'169;
                 E   =>   'd'196;
                 F   =>   'd'225;
        }
```

Fig. 6-7 CUPL logic description file for example 6-2

A possible CUPL logic description file solution is shown in Fig. 6-7. This solution uses a truth table design entry technique to define the desired function. The 4 input bits have been grouped together and named "data_in," while the 8 output bits are grouped under the name "square" using the field declarations. A truth table format is declared using the keyword "table" followed by the name of the input bit field, the table assignment symbol "=>", the name of the output bit field, and then the list of input and output assignments, which is enclosed in braces { }. The input bits ("data_in") are given in hexadecimal (the default number base) and the output bits ("square") are given in decimal (specified with the prefix 'd').

Errors are reported on-screen in the Message window during compilation of the logic description file and will also be logged in the ".LST" file if the "list" Output Format option is selected under Compiler Options. If, for example, the following error was made in the output pin assignment statement of the logic description file:

```
Pin [12..18] = [N0..7];   /* Square Output Value */
```

then the resultant SQUARE.LST file would be as shown in Fig. 6-8. The error message is indicating that a total of 7 pins have been assigned to 8 variable names, which is a mismatch in size. The error pointer is pointing to the end of the line where the error occurred.

Example 6-3

Implement a 4-channel multiplexer using a GAL16V8. The logic expression for this circuit is:

$$Y = (D0\ \overline{SELB}\ \overline{SELA}\ +\ D1\ \overline{SELB}\ SELA$$
$$+\ D2\ SELB\ \overline{SELA}\ +\ D3\ SELB\ SELA)\ \overline{EN}$$

An example CUPL logic description file to implement this combinational circuit is given in Fig. 6-9. The active-low input signal level for EN has been specified in the pin assignment statement, while the logic equation indicates that the EN signal has to be asserted (or true) to produce the desired logic function. List notation has been utilized to assign pins 2 through 5, respectively, to input variables D0 through D3. The "D" variable specifications are accomplished using CUPL's indexed variable notation. The input variables SELA and SELB are also assigned to pins 6 and 7, respectively, using the list notation technique. The list notation provides a very convenient shortcut. This example also illustrates the use of intermediate variable definitions. Each of the four possible combinations of the two select controls has been assigned a variable name to make the writing of the logic equation less complex. An easy way to interpret the logic expression is to substitute the word "if" for the ANDs in the equation.

```
"Y is equal to (D0 if SEL0 is true or D1 if SEL1
is true or D2 if SEL2 is true or D3 if SEL3 is
true) only if the circuit is enabled."
```

```
LISTING FOR LOGIC DESCRIPTION FILE: SQUARE.pld      Page 1
CUPL(WM): Universal Compiler for Programmable Logic
Version 4.7a Serial# MW-66999998
Copyright (c) 1983, 1996 Logical Devices, Inc.
Created Mon Dec 30 14:38:15 1996

   1:Name        SQUARE;
   2:Partno      L105-2;
   3:Date        02/04/91;
   4:Revision    03;
   5:Designer    Greg Moss;
   6:Company     Digi-Lab, Inc.;
   7:Assembly    Generator board;
   8:Location    U210;
   9:Device      G16V8A;
  10:Format      j;
  11:
  12:/***********************************************/
  13:/* Generates the square of an input value using */
  14:/* the truth table design entry technique.     */
  15:/***********************************************/
  16:
  17:/**   Inputs  **/
  18:Pin [2..5] = [I3..0];     /* Data Input Value    */
  19:
  20:/**   Outputs  **/
  21:Pin [12..18] = [N0..7];  /* Square Output Value */

[0012ca] vector size mismatch:  lhs size = 7, rhs size = 8
  22:
  23:/** Declarations and Intermediate
  24:                        Variable Definitions **/
  25:field data_in = [I3..0];
  26:field square = [N7..0];
  27:
  28:/** Logic Description  --  truth table format **/
  29:table  data_in  =>  square   {
  30:              0   =>  'd'0;
  31:              1   =>  'd'1;
  32:              2   =>  'd'4;
  33:              3   =>  'd'9;
  34:              4   =>  'd'16;
  35:              5   =>  'd'25;
  36:              6   =>  'd'36;
  37:              7   =>  'd'49;
  38:              8   =>  'd'64;
  39:              9   =>  'd'81;
  40:              A   =>  'd'100;
  41:              B   =>  'd'121;
  42:              C   =>  'd'144;
  43:              D   =>  'd'169;
  44:              E   =>  'd'196;
  45:              F   =>  'd'225;
  46:         }
  47:
```

Fig. 6-8 CUPL Listing file for error in example 6-2

```
        Name        4CH_MUX;
        Partno      L105-3;
        Date        06/29/93;
        Revision    01;
        Designer    Greg Moss;
        Company     Digi-Lab, Inc.;
        Assembly    Multiplexer Board;
        Location    U315;
        Device      G16V8A;
        Format      j;

/******************************************************/
/*    PLD design for a 4-input multiplexer circuit    */
/******************************************************/

/**   Input Pin Assignments   **/
Pin   1   = !EN;        /* Enable control - active low  */
Pin [2..5] = [D0..3];   /* Input data (list)            */
Pin [6,7] = [SELA,SELB];    /* Select controls (list) */

/**   Output Pin Assignments   **/
Pin  19   = Y;          /* Multiplexer output           */

/**   Declarations & Intermediate Variable Definitions  **/
/* Defining select combinations for:    SELB  SELA   */
SEL0 = !SELB & !SELA;                 /*  0     0    */
SEL1 = !SELB &  SELA;                 /*  0     1    */
SEL2 =  SELB & !SELA;                 /*  1     0    */
SEL3 =  SELB &  SELA;                 /*  1     1    */

/**   Logic Equations   **/
Y   = ( D0   &   SEL0
    #     D1  &   SEL1
    #     D2  &   SEL2
    #     D3  &   SEL3 )  &  EN;
```

Fig. 6-9 CUPL logic description file for example 6-3

Example 6-4

Design a logic circuit that will add two, 2-bit numbers and output the 3-bit sum.

An example CUPL logic description file to implement the adder circuit is given in Fig. 6-10. The two, 2-bit operands (A3 A2 and A1 A0) have been named as one indexed variable to make it easy to group the two input numbers in a field statement. A truth table was used to define the logic function. The input bits ("OPERANDS") are given in binary and the output bits ("SUM") are given in decimal. The comments are used to document the arithmetic in the table.

```
Name       adder;                Partno    L159-6;
Date       12/08/96;             Rev       01;
Designer   Greg Moss;            Company   Digi-Lab, Inc.;
Assembly   Arith Board;          Location  U506;
Device     G16V8A;               Format    j;

/**************************************************/
/* 2 bit adder using a truth table               */
/**************************************************/

/** Inputs **/
Pin [1..4] = [A3..2,A1..0];      /* two, 2-bit operands */

/** Outputs **/
Pin [12..14] = [S0..2];          /* 3-bit sum           */

/** Declarations & Intermediate Variable Definitions **/
field OPERANDS = [A3..0];        /* operands            */
field SUM = [S2..0];             /* sum                 */

/** Truth Table  **/
table   OPERANDS => SUM     {
        'b'0000  => 'd'0;        /* 0 + 0 = 0 */
        'b'0001  => 'd'1;        /* 0 + 1 = 1 */
        'b'0010  => 'd'2;        /* 0 + 2 = 2 */
        'b'0011  => 'd'3;        /* 0 + 3 = 3 */
        'b'0100  => 'd'1;        /* 1 + 0 = 1 */
        'b'0101  => 'd'2;        /* 1 + 1 = 2 */
        'b'0110  => 'd'3;        /* 1 + 2 = 3 */
        'b'0111  => 'd'4;        /* 1 + 3 = 4 */
        'b'1000  => 'd'2;        /* 2 + 0 = 2 */
        'b'1001  => 'd'3;        /* 2 + 1 = 3 */
        'b'1010  => 'd'4;        /* 2 + 2 = 4 */
        'b'1011  => 'd'5;        /* 2 + 3 = 5 */
        'b'1100  => 'd'3;        /* 3 + 0 = 3 */
        'b'1101  => 'd'4;        /* 3 + 1 = 4 */
        'b'1110  => 'd'5;        /* 3 + 2 = 5 */
        'b'1111  => 'd'6;   }    /* 3 + 3 = 6 */
```

Fig. 6-10 CUPL logic description file for example 6-4

Laboratory Projects

Design PLD logic circuits for the following applications. Use a text editor to create (and edit) the design's logic description file. Compile the logic description file to produce a JEDEC output file. Program a GAL16V8 with the JEDEC file and test each of your circuit designs in the lab.

6.1   Pattern generator
Design a logic circuit that will output the pattern illustrated by the following timing diagram. The inputs are DCBA and the outputs are WXYZ.

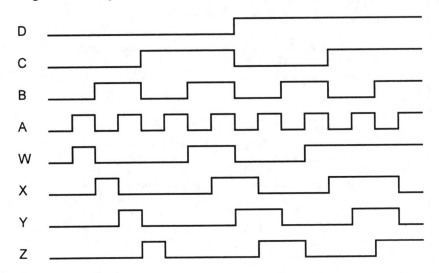

6.2   2421-to-5421-BCD code converter

Design and construct a 2421-BCD-to-5421-BCD code converter. The truth table for this design is given below. The inputs are labeled DCBA and the outputs are WXYZ. Note that in this situation, we care about only 10 of the 16 possible input combinations. The other 6 input combinations are listed at the bottom of the truth table and are labeled as "invalid." The output for each of these input conditions is given a default output value of 1111.

| Decimal | 2 | 4 | 2 | 1 | 5 | 4 | 2 | 1 |
Value	D	C	B	A	W	X	Y	Z
0	0	0	0	0	0	0	0	0
1	0	0	0	1	0	0	0	1
2	0	0	1	0	0	0	1	0
3	0	0	1	1	0	0	1	1
4	0	1	0	0	0	1	0	0
5	1	0	1	1	1	0	0	0
6	1	1	0	0	1	0	0	1
7	1	1	0	1	1	0	1	0
8	1	1	1	0	1	0	1	1
9	1	1	1	1	1	1	0	0
invalid	0	1	0	1	1	1	1	1
invalid	0	1	1	0	1	1	1	1
invalid	0	1	1	1	1	1	1	1
invalid	1	0	0	0	1	1	1	1
invalid	1	0	0	1	1	1	1	1
invalid	1	0	1	0	1	1	1	1

6.3   Twos complementer

Design a logic circuit that will produce a 5-bit output ($X4$ $X3$ $X2$ $X1$ $X0$) that is equal to the two's complement of a 5-bit input value ($I4$ $I3$ $I2$ $I1$ $I0$).

6.4   BCD-to-binary converter

Design a logic circuit that will convert a 5-bit BCD input into its equivalent 4-bit binary value. Since the output is only 4 bits long, the largest number that is to be converted is the decimal value 15. This converter circuit should also have an output called **ERR** that will be high if the BCD input value is not in the range of 0 through 15. Hints: Use the truth table design entry technique to define valid input conditions and resultant outputs. Write a logic equation using the equality operator to detect the ranges of input values that should produce a high output for ERR.

6.5   Programmable logic unit

Design a programmable logic circuit that will perform the logic operations given in the table below on the two 4-bit inputs. X and Y are controls that determine which of the four functions is to be performed by the logic circuit. A and B represent two different 4-bit data inputs and F represents the resultant 4-bit data output that will be produced. Each of the output bits will be a function of corresponding A-bit and B-bit and the X and Y controls. For example, F3 is dependent on A3, B3, X, and Y, while F2 is dependent on A2, B2, X, and Y. Hint: The easiest solution is to write a <u>set</u> of four (one for each F output) logic equations in the following form (where n is 0 through 3):

```
Fn  =   (An # Bn)   &   (!X & !Y)
    #   (An & Bn)   &   (!X &  Y)
    #   (An $ Bn)   &   ( X & !Y)
    #     (!An)     &   ( X &  Y);
```

```
X   Y │ Operation
0   0 │ F = A # B
0   1 │ F = A & B
1   0 │ F = A $ B
1   1 │ F = !A
```

```
Data inputs:
      A = (A3 A2 A1 A0)
      B = (B3 B2 B1 B0)
Data output:
      F = (F3 F2 F1 F0)
```

6.6   Binary-to-BCD converter

Design a logic circuit that will convert a 5-bit binary input into its equivalent 2-digit (only 6 output bits are needed) BCD value. The converter can handle numbers from 0_{10} through 31_{10}.

6.7   Binary multiplier

Design a logic circuit that will multiply two binary numbers together. The multiplicand is a 3-bit number and the multiplier is a 2-bit number. The product will be a 5-bit number.

6.8   Data switcher
        Design a programmable logic circuit that will route two input data bits to the selected
        outputs as given in the function table below. S1 & S0 are controls that determine
        which of the two inputs (A & B) is to be routed to each of the two outputs (Y1 & Y0).
        A & B and Y1 & Y2 are each single bits. EN is an active-low enable for the data
        switcher. If the enable is high, we don't care what input levels are on S1 or S0. Hint:
        The easiest solution is to write equations for each of the outputs Y1 & Y0. Declare a
        field name for S1 & S0 use the equality operator to determine when each output should
        be equal to A and when each output should be equal to B. Don't forget the active-low
        enable.

EN	S1	S0	Y1	Y0
0	0	0	A	B
0	0	1	A	A
0	1	0	B	B
0	1	1	B	A
1	X	X	0	0

ANALYZING FLIP-FLOP AND LATCH CIRCUITS

Objectives

- To test SR and D latches and JK flip-flops.
- To debounce a toggle switch using a simple SR latch.
- To test latch and flip-flop applications in parallel data latch and binary counter circuits.
- To measure the waveforms generated by a binary counter using an oscilloscope.

Suggested Parts				
74LS00	74LS75	74LS112A	74LS393	SPDT Switch
1 KΩ Resistors				

Flip-Flops and Latches

Combinational logic circuits have outputs that are dependent only on the current inputs to the circuit. Sequential circuits, on the other hand, are dependent not only on the current inputs but also on the prior circuit conditions. Sequential circuits contain memory elements that allow them to utilize the prior circuit conditions in determining the circuit output. These memory elements consist of flip-flops and latches. There are various categories of flip-flops and latches including SR, D, and JK types.

Several flip-flops or latches can be connected together in specific circuit configurations to construct various types of registers and counters. The 74LS75 is an example of a register chip. It is called a 4-bit latch since it contains four D flip-flops that can be used to store a 4-bit number. The 74LS393 contains two separate 4-bit binary counters. Each counter has a triggering (or clock) input that makes it count up one in binary for each pulse input (negative-edge) received. The counters also have an active-high reset (or clear) input that forces the counter to zero and holds it there as long as the reset is asserted.

Example 7-1

Analyze the operation of the unclocked SR latch circuit shown in Fig. 7-1 when the given input waveforms are applied.

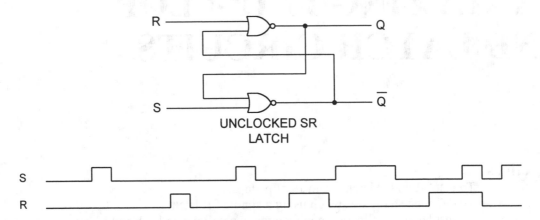

Fig. 7-1 Unclocked SR latch and input conditions

The SR latch shown in Fig. 7-1 will function as indicated in Table 7-1. The specified input waveforms will produce the results shown in Fig. 7-2. The section of the output waveforms that is shown in a cross-hatch cannot be determined. The section which is marked with an asterisk is the result of applying an invalid input condition (S = R = 1).

S	R	Q	$\overline{Q}$	Command
0	0	Q	$\overline{Q}$	no change
0	1	0	1	reset
1	0	1	0	set
1	1	0	0	invalid

Table 7-1 Truth table for active-high input SR latch

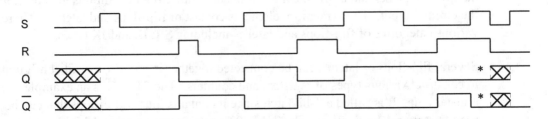

Fig. 7-2 Timing diagram results for example 7-1

Example 7-2

Analyze the register circuit shown in Fig. 7-3 for the given input waveforms. Assume that the initial condition is Q3 Q2 Q1 Q0 = 1 0 1 0. The results are shown in Fig. 7-4.

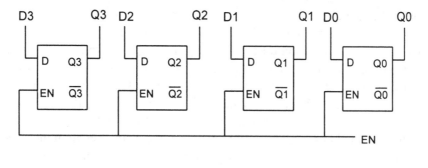

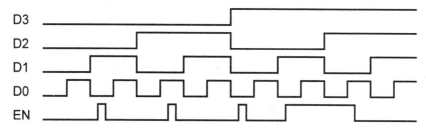

Fig. 7-3 4-bit transparent register and input conditions for example 7-2

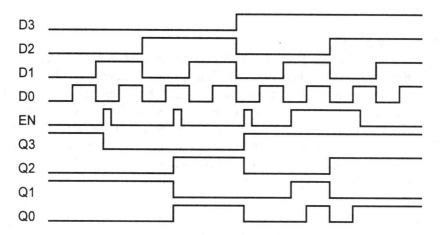

Fig. 7-4 Timing diagram for example 7-2

Laboratory Projects

7.1 NAND SR latch

Construct and test a NAND SR latch. Use two logic switches for the control inputs (Set and Reset) and monitor both latch outputs (Q and Qbar) with separate lamps. Test the latch under each of the four possible input conditions. How do you set the latch? What is the condition of the latch's two outputs (Q and Qbar) when the latch is set? How do you reset the latch? What is the condition of the latch's outputs when the latch is reset? How do you control the latch so that it will hold a bit of data? Why are the outputs named Q and Qbar? What happens to a NAND latch in the invalid state? Are the latch's control inputs active-high or active-low? Describe the operation of the latch.

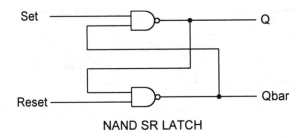

NAND SR LATCH

7.2 Debouncing a logic switch

(a) Try to manually clock a 74LS393 binary counter with a simple digital switch (see the following schematic). The jumper wire connects the circuit node labeled "Simple Digital Switch Output" to the CLK (clock) pin on the counter chip. Monitor the clocking signal with a logic probe as shown in the drawing. Ideally, the lamps connected to the output of the counter chip **should** display a binary sequence that increments by one count each time the logic switch is flipped back and forth. Does the counter seem to count correctly? What is the counter actually doing each time the switch is cycled? What is switch bounce? How does switch bounce affect a digital circuit?

(b) Move the jumper wire for the 74LS393 clock to the output of the NAND SR latch ("Debounced Digital Switch Output") to eliminate the switch's contact bounce and try clocking the counter again. Does the counter now operate correctly? What is the sequence of states for the counter's outputs? Describe the operation of the debouncing circuit. How does the latch eliminate the contact bounce of the switch? Disconnect the two pull-up resistors connected to the switch and determine the effect on the circuits operation, if any. Why should the pull-up resistors be used?

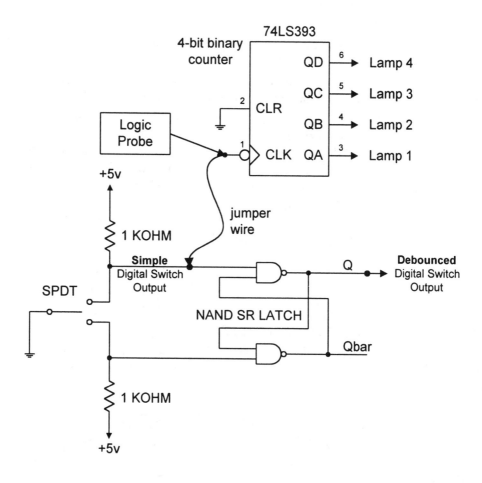

7.3 D latch

Construct and test a D latch. Use a logic switch for the Data input and another for the Enable. Monitor both latch outputs (Q and Qbar) with separate lamps. What does the latch do when the enable is high and the data input is changed? What does the latch do when the enable is low and the data input is changed? Is the latch's Enable active-high or active-low? How do you store data in the D latch? What part of the D latch circuit actually stores the data and which part provides the data steering (or control) of the latch? Describe the operation of the latch.

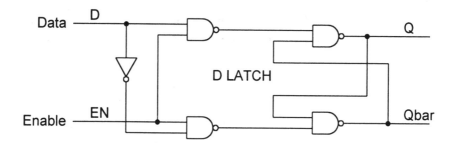

7.4   4-bit D latch

Use a low frequency TTL pulse (approximately 1 Hertz) to automatically clock the 4-bit binary counter in a 74LS393. Connect the 4-bit latch circuit (4-bit parallel shift register) in a 74LS75 to the output of the counter as shown below. Monitor the counter's output with 4 logic lamps and monitor the quad latch's outputs with 4 additional lamps. Test the operation of the latch by clocking the counter while the latch enables are held high and then low. Which of the latch outputs is the LSB? Which is the MSB? Why? Describe the circuit's operation. What type of latch is contained in the 74LS75? The latch is enabled when it is picking up new input data and passing it through to the outputs. The latch is disabled when it is storing data regardless of the signals applied to the inputs. Are the latch enables active-high or active-low? Is level triggering or edge triggering used in enabling the latches? Why is this 4-bit latch called a "transparent" latch?

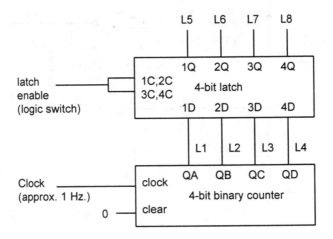

Note: L1 through L8 represent 8 lamp monitors

7.5   JK flip-flop

Test the operation (both synchronous and asynchronous) of one of the JK flip-flops contained in a 74LS112A. Connect the flip-flop's J, K, PRE, and CLR control inputs to logic switches, the clock to a pushbutton, and the outputs (Q and Qbar) to lamps. Which control inputs are synchronous? Determine how to synchronously control the flip-flop to hold data, set, reset, and toggle. Which control inputs are asynchronous? Determine how to asynchronously store a zero or a one in the flip-flop. Why are the controls classified as synchronous or asynchronous? Which set of controls (synchronous or asynchronous) have priority? What is the function of the clock?

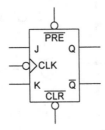

7.6   Binary counter
Construct and test the following 3-bit binary counter. Use a pushbutton to manually
clock the counter and observe the counter's output (QC QB QA) on 3 lamps. Assume
that QC is the most significant bit and QA is the least significant bit. What is the count
sequence produced by this circuit? Is this an up-counter or a down-counter? What is
the counter's modulus?

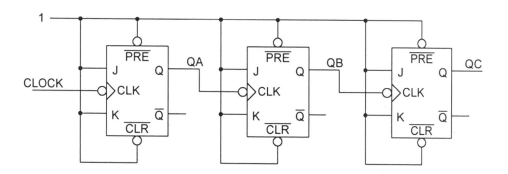

7.7   Counter timing diagram
Use an oscilloscope to observe the input and output waveforms (4 signals: clock, QA,
QB, QC) for the 3-bit counter in Project 7.6. A good display on the oscilloscope will
be obtained if the clock frequency is increased to at least 10 KHz. Remember that the
clock signal must be TTL compatible. A 4-channel scope will provide the best display
since all 4 waveforms can be viewed simultaneously, but a dual-channel scope can also
be used by carefully swapping signals. Arrange the scope display in the order of
decreasing signal frequency with the highest frequency signal at the top of the screen
and the lowest frequency signal at the bottom. Trigger the oscilloscope on the lowest
frequency signal. Which signal appears to be the lowest frequency? Why? Carefully
draw the waveforms to show the timing relationships between each of the counter
outputs and the clock signal.

UNIT 8

TIMING AND WAVESHAPING CIRCUITS

Objectives

- To convert analog signals into logic-compatible signals using Schmitt trigger devices.
- To design and construct astable multivibrators that produce specified square waveforms.
- To design and construct monostable multivibrators that produce specified time delay patterns.

Suggested Parts			
74LS04	74LS14	74LS221	555
1N4001			
Resistors:	1.0, 2.2, 27, 33, 47, 62, 68, 82 KΩ		
Potentiometer:	10KΩ (10-turn)		
Capacitors:	0.001, 0.0047, 0.01, 0.1, 10 µfarads		

Schmitt Triggers

If a slow changing signal is applied to the input of a logic device, the device will often produce an output signal that oscillates as the input slowly transitions between high and low logic levels. A Schmitt trigger is a waveshaping device used to convert signals in which the voltage is slowly changing (and, therefore, incompatible with logic devices) to a signal of the same frequency but is compatible with the signal transition times of logic devices. Schmitt trigger devices have an input hysteresis since they have two different specific triggering points. A positive-going threshold voltage and a negative-going threshold voltage will switch the Schmitt trigger's output between the two possible logic levels. The 74LS14 IC is a hex inverter chip with Schmitt trigger inputs.

One-Shots or Monostable Multivibrators

A one-shot is a timing device in which the output is triggered into a quasi-stable state and then returns to its stable state. The length of the quasi-stable state is usually controlled by an external resistor and capacitor. One-shots are typically used to produce delays in control signals for digital systems. There are two types of one-shots: retriggerable and nonretriggerable. The 74LS221 IC contains two independent, nonretriggerable one-shots.

Clocks or Astable Multivibrators

An astable or free-running multivibrator has an output that continually switches back and forth between the two states. This type of signal is often used as a clock signal to control (or trigger) synchronous circuits. The 555 IC timer is a device that can be used to produce a clock signal whose frequency and duty cycle are dependent on external timing resistors and capacitors. The 555 can also be configured as a one-shot.

Laboratory Projects

Design timing and waveshaping circuits for the following applications. Breadboard and verify your designs.

8.1   Schmitt trigger waveshaper
Use an oscilloscope to compare the output waveforms produced by a NOT gate in a 74LS04 and a 74LS14 (Schmitt trigger NOT) when either a triangle or sine waveform from a signal or function generator is applied to the inputs of the gates. **Be sure to adjust the generator signal using the oscilloscope to $0 \leq V_i \leq +5v$ before applying it to the gate inputs.**

8.2   Pulse stretcher
Construct a pulse stretcher circuit using a 74LS221 one-shot. Select appropriate components to make the pulse width approximately 0.5 sec. How can this circuit also be used to eliminate the contact bounce problem of mechanical switches?

8.3   Variable frequency clock
        Design and construct a clock generator circuit using a 555 timer. The duty cycle of the
        clock waveform should be approximately 50 percent. The clock output frequency
        should be variable in steps (by changing the 555 timing capacitor value). The 3 clock
        frequency values should be approximately 1 Hz, 1 KHz, and 10 KHz.

8.4   Waveform generator
        Design and construct a waveform generator that will produce the waveforms given in
        the timing diagram below. Use the 555 timer to generate one of the waveforms, and
        then use that waveform to trigger the 2 one-shots in the 74LS221 to produce the other
        two waveforms.

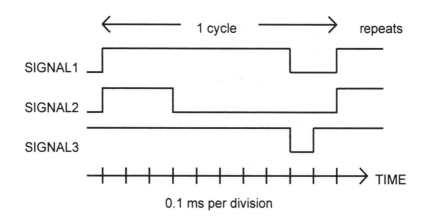

8.5 Variable duty-cycle square wave generator

Design a variable duty-cycle square wave generator using a 555 timer. The output frequency should be 10 KHz and the duty-cycle should vary from 15% to 85%. Use a 10-turn 10KΩ potentiometer to control the duty cycle. The diode in the modified astable multivibrator circuit below allows the R_B resistance to be shorted out during the charge cycle. The new equations for this circuit will be:

$$t_L = 0.693 \ R_B \ C$$
$$t_H = 0.693 \ R_A \ C$$

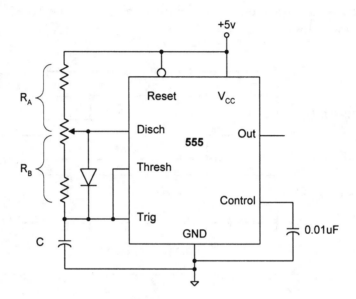

8.6 Delayed pulse

Design a timing circuit that will output a positive pulse that is 0.22ms wide. The output pulse is delayed and does not start until 0.33ms after being triggered by a negative-edge signal.

8.7 Clock generator using one-shots

Design a 100 Hz clock generator using the two one-shots contained in a 74LS221. Use a 0.1 μF timing capacitor for each one-shot. Select a timing resistor for each one-shot. Hint: Let the time out of each one-shot trigger the other.

ARITHMETIC CIRCUITS

Objective

- To apply parallel adder circuits in arithmetic applications.

Suggested Parts					
74LS04	74LS08	74LS32	74LS83A	74LS86	GAL16V8

Adder Circuit Applications

The primary building blocks in adder circuits are half adders and full adders. The basic difference between half and full adders is that a full adder has an additional input (3 inputs total) that allows a carry input to be handled. Both half and full adders generate a sum and a carry output. These building blocks can be implemented a number of ways using various logic gates.

Due to their versatility and usefulness, parallel adders made up of several full adder stages are available as integrated circuit devices. The 74LS83 shown in Fig. 9-1 (or 74LS283) is an example of a 4-bit parallel binary adder chip. This MSI chip can add two 4-bit binary numbers (A4 A3 A2 A1 and B4 B3 B2 B1) together. The adder chip outputs a 5-bit sum (C4 Σ4 Σ3 Σ2 Σ1). There is also a carry input (C0) available that can be used to connect multiple 74LS83 chips together for larger adder applications.

An 8-bit parallel adder constructed with two 74LS83 chips is illustrated in Fig 9-2. The inputs A8 through A1 and B8 through B1 are added together to produce the 9-bit sum labeled S9 through S1. Parallel binary adders can also be used in many arithmetic applications besides just simple addition.

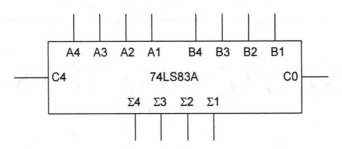

Fig. 9-1 74LS83, 4-bit parallel binary adder chip

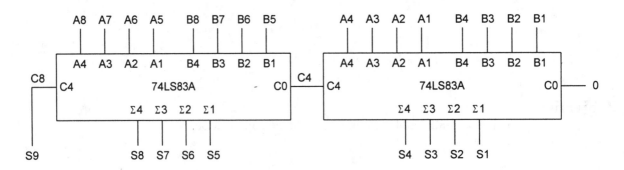

Fig. 9-2 8-bit parallel binary adder circuit using two 74LS83 chips

Example 9-1

Design a 2-bit adder using a GAL16V8.

An example solution is shown in Fig. 9-3. This example illustrates applying a user-defined function in a CUPL design. The keyword "function" is followed by the name of the function (FA is the function's name in this example). This name can then be used in a logic equation to represent the function. Enclosed in parenthesis, after the function name, is a list of input and output parameters for the function. The body of the function is enclosed in braces {} and may contain any combination of logic equations, truth tables, state-machine syntax, etc. A function must be defined before it can be used. This design uses the function in two different equations with different input and output parameters applied in each case.

```
Name      2-bitadd;    Partno    L159-5;
Date      12/08/96;    Rev       01;
Designer  Greg Moss;   Company   Digi-Lab, Inc.;
Assembly  Adder Board; Location  U505;
Device    G16V8A; Format    j;

/*******************************************************/
/* 2 bit adder using the CUPL function statement.      */
/*******************************************************/

/** Inputs **/
Pin [1..2] = [A2..1];              /* First 2-bit number  */
Pin [3..4] = [B2..1];              /* Second 2-bit number */

/** Outputs **/
Pin [18..19] = [S2..1];            /* 2-bit sum           */
Pin [12..13] = [C1..2];            /* Carries             */

/** Define full adder function **/
function FA(A, B, Cin, Cout) {     /* (I/O parameters)    */
   Cout = Cin & (A $ B) # A & B;   /* Compute carry out */
   FA   = Cin $ (A $ B);      }    /* Compute sum         */

/** Equations **/
S1 = FA(A1, B1, 'b'0, C1);   /* Initial carry in = 'b'0 */
S2 = FA(A2, B2,  C1, C2);    /* Carry out = C2          */
```

Fig. 9-3 2-bit parallel binary adder using a CUPL function statement

Laboratory Projects

Design arithmetic circuits for the following applications. Breadboard and verify your designs.

9.1   Two-bit adder

Design a 2-bit binary adder with the 74LS83A. A 2-bit adder adds two 2-bit numbers together and produces a 3-bit sum. **Note that the unused inputs should be connected to some appropriate logic level and not left floating.** There are several valid ways to connect this chip for this application. Compare the number of chips needed for this design with one that uses only logic gates for the circuit implementation.

```
        A1   A0
    +   B1   B0
   _____
   S2   S1   S0
```

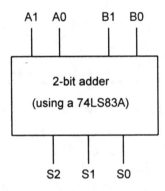

9.2   Excess-3/8421 BCD code converter

Design a code converter circuit using a 74LS83A and any additional necessary logic gates that will convert either an excess-3 coded input value to an equivalent 8421 BCD coded output or an 8421 BCD coded input value to an equivalent excess-3 coded output. The conversion function will be controlled by F, as shown in the following function table.

F	Function
0	8421 to X-3
1	X-3 to 8421

9.3   Selected arithmetic function generator

Design a logic circuit using the 74LS83A and any other necessary gates that will
perform the following fixed arithmetic operations on a 4-bit input number (I3 I2 I1 I0).
M and N are control inputs to the circuit that determine which of the 4 arithmetic
operations is to be performed. The 4-bit output is F3 F2 F1 F0. Test a representative
sampling of data for this circuit.

$$I = (I3\ I2\ I1\ I0)$$
$$F = (F3\ F2\ F1\ F0)$$

M	N	Operation
0	0	F = I + 1
0	1	F = I - 1
1	0	F = I + 2
1	1	F = I - 2

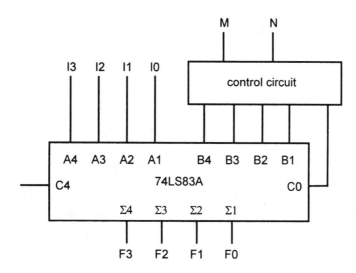

9.4   Four-bit adder/subtractor

Design a 4-bit adder/subtractor using the 74LS83A and any other necessary logic gates (or a PLD for both the controlled complement block and the overflow block). The adder/subtractor circuit can handle signed numbers (using two's complement arithmetic techniques). F is a control input to the circuit that determines if the circuit will add
(A + B) or subtract (A - B) the 4-bit data inputs. The overflow detector circuit will output a high signal if the adder/subtractor produces an incorrect result due to an overflow condition. Overflow occurs when like-signed numbers are added together but the sign of the result is not the same sign. Hints: Perform subtraction by forming the one's complement of the B input (the subtrahend) and adding 1 (with the carry input) plus the A input (the minuend). Remember that the B input should not be complemented for addition (F = 0). To design the overflow circuit, define in a truth table when an overflow exists by monitoring each of the 3 sign bits (2 operands and answer).

F	Function	
0	A+B	(addition)
1	A-B	(subtraction)

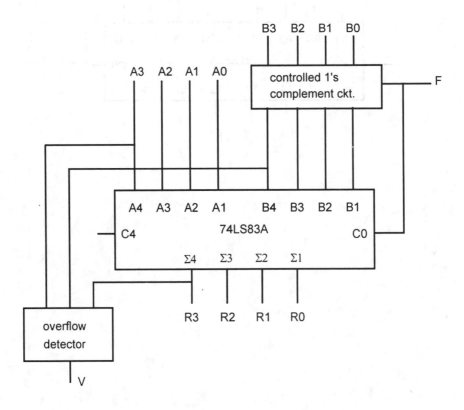

9.5   Binary multiplier

Design an unsigned binary multiplier using the 74LS83A and two 74LS08 chips. The multiplier circuit should handle 4-bit multiplicands (A3 A2 A1 A0) and 2-bit multipliers (B1 B0) to produce 6-bit products (P5 P4 P3 P2 P1 P0). Hint: Use the AND gates to multiply <u>each pair</u> of bits.

		A3	A2	A1	A0	multiplicand
				B1	B0	multiplier
		$A3 \cdot B0$	$A2 \cdot B0$	$A1 \cdot B0$	$A0 \cdot B0$	partial-product
	$A3 \cdot B1$	$A2 \cdot B1$	$A1 \cdot B1$	$A0 \cdot B1$		partial product
P5	P4	P3	P2	P1	P0	product

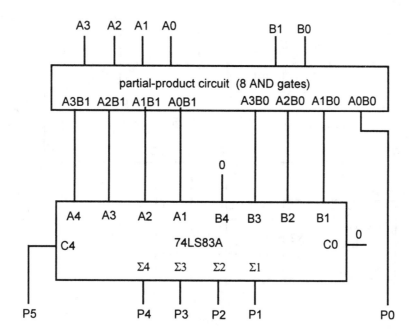

9.6   4-bit binary adder

Design a 4-bit binary adder using a GAL16V8. The adder should handle two, 4-bit operands plus a carry input to produce a 4-bit sum and a carry output. Note: Pins 15 and 16 on the GAL16V8 do not produce internal feedback in the simple programming mode.

9.7 BCD adder

Design a 4-bit BCD adder using the 74LS83A and a GAL16V8. The two BCD inputs are P3 P2 P1 P0 and Q3 Q2 Q1 Q0. Cin is the carry input for the BCD adder stage. The BCD digit sum is S3 S2 S1 S0, and Cout is the BCD carry output. Hints: The parallel adder chip performs the initial binary addition of the 4-bit numbers and the GAL chip detects if a BCD carry output (Cout) should be produced and, if so, will also correct the binary sum from the 74LS83A. Cout will be high if the 5-bit binary sum produced by the 74LS83A is greater than nine. If a carry out is produced, the binary sum must be corrected by adding six (0110_2) to the lower four bits to create the proper BCD digit. If the carry out is zero, no correction is necessary (i.e., add 0000). The sum correction block can be simplified to an adder circuit, which consists of two half adders (to produce S1 and S3) and one full adder (to produce S2). Note: Pins 15 and 16 on the GAL16V8 do not produce internal feedback in the simple programming mode.

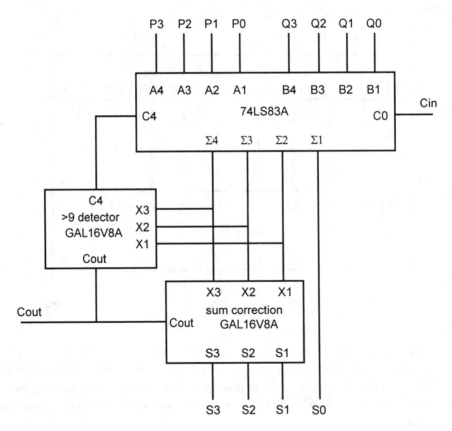

ASYNCHRONOUS COUNTERS

Objectives

- To analyze asynchronous counter circuits to predict their theoretical operation.
- To construct and test asynchronous counter circuit designs.

Suggested Parts				
74LS00	74LS08	74LS10	74LS20	74LS112A

Asynchronous Counters

Asynchronous or ripple counters are triggered by separate clocking signals on each flip-flop. Output signals from appropriate flip-flops are used to provide the triggering signals for other flip-flops. Because of this clocking arrangement, the individual flip-flops do not change states at the same time. Determining the count sequence of an asynchronous counter requires that each flip-flop be individually analyzed, starting with the one triggered by the system clock. If a flip-flop receives the proper (depends on whether the flip-flops are positive- or negative-edge triggered) triggering signal, its resultant output will depend on the logic levels present on the flip-flop's synchronous inputs. The count sequence may be an up- or down-count depending on the circuit's gating controls and the triggering signals used. Additionally, the flip-flop asynchronous control inputs, preset and clear, may be used to modify the count sequence. The mod number of a counter is the number of states present in the count sequence.

Example 10-1

Analyze the asynchronous counter given in Fig. 10-1 and determine its sequence of states and modulus. Sketch the circuit's timing diagram.

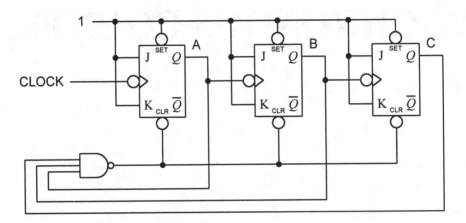

Fig. 10-1 Schematic for asynchronous counter in example 10-1

A simple approach to the analysis of this circuit would be to note that the circuit uses a common, asynchronous counter design. The J and K inputs for each flip-flop are all tied high. This will result in a toggle function each time the flip-flop is clocked by the negative-edge of the individual clocking signals for the asynchronous counter. The JK flip-flops are each triggered by the Q output of the previous flip-flop. This clocking arrangement will produce a recycling, binary, up-count sequence until an asynchronous flip-flop control input on PRE or CLR is activated. The PRE and CLR inputs are active-low. In this circuit the PRE inputs are all inactive and will not affect the count sequence since they are tied high. The CLR inputs, on the other hand, are all controlled by the output of the NAND gate. This gate is called a decoding gate since it will decode or detect the counter state 111. When the counter reaches the decoded state of CBA = 111, the output of the NAND will go low, which in turn should cause the counter to clear (CBA = 000). The theoretical counter sequence is given in Table 10-1. The counter state 111 (shown underlined in the table) is called a transient state since the counter will immediately clear instead of waiting for the next clock pulse to recycle. The counter is a mod-7 counter since it is effectively skipping one state in the binary count sequence. The regular states in the count sequence for this counter are 0 through 6. Transient states are not counted when determining the modulus of a counter. The timing diagram for this counter is drawn in Fig. 10-2. Notice the glitch (spike) that is produced in the flip-flop A output waveform. This spike is caused by the transient state (111) of the counter. Flip-flop A goes from a 0 during state 110 to a 1 for the transient state, and then back to a 0 when the counter is cleared by the decoding gate. The transient state exists only for a very short time (typically, only a few nanoseconds).

CLOCK ↓	C	B	A	
0	0	0	0	
1	0	0	1	
2	0	1	0	
3	0	1	1	
4	1	0	0	
5	1	0	1	
6	1	1	0	
7	1	1	1	*transient*
	0	0	0	
8	0	0	1	
9	0	1	0	

Table 10-1 Sequence for asynchronous counter in example 10-1

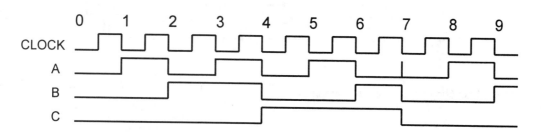

Fig. 10-2 Timing diagram for asynchronous counter in example 10-1

Laboratory Projects

Analyze the following asynchronous counters. For each counter, determine the counter's modulus and sketch its timing diagram (show the clock input and the flip-flops' Q outputs). Construct and test the operation of the counter circuits. Use an oscilloscope to compare the output waveforms for each of the recycling counters with your theoretical prediction.

10.1   Recycling asynchronous counter 1

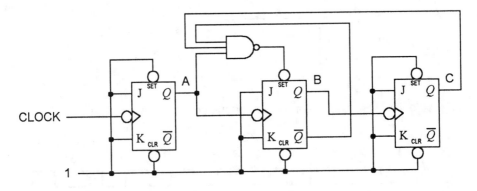

10.2   Recycling asynchronous counter 2

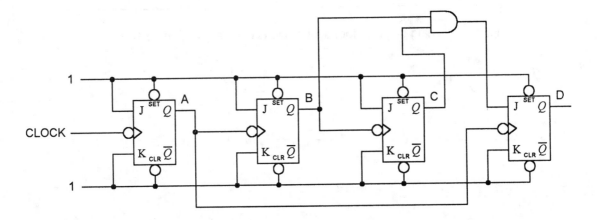

10.3 Recycling asynchronous counter 3

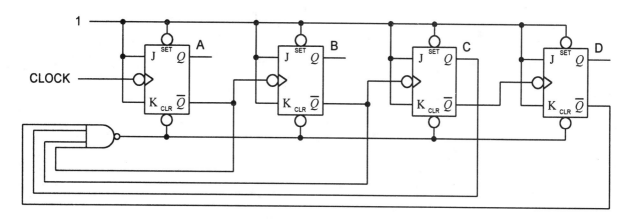

10.4 Recycling asynchronous counter 4

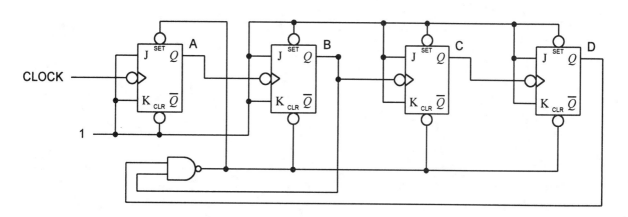

10.5 Self-stopping asynchronous counter

Determine the count sequence produced after an active-low RESTART pulse has been applied to the counter.

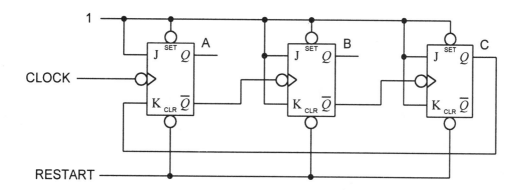

SYNCHRONOUS COUNTERS

Objectives

- To analyze synchronous counter circuits to predict their theoretical operation.
- To construct and test synchronous counter circuit designs.

Suggested Parts				
74LS00	74LS08	74LS32	74LS20	74LS112A

Synchronous Counters

Synchronous or parallel counters are triggered by a common clocking signal applied to each flip-flop. Because of this clocking arrangement, all flip-flops react to their individual synchronous control inputs at the same time. The count sequence depends on the control signals input to each flip-flop. Additionally, the flip-flop asynchronous control inputs, preset and clear, may be used to modify the count sequence.

Example 11-1

Analyze the synchronous counter circuit given in Fig. 11-1. Draw the state transition diagram (include <u>all</u> 8 possible states) for the counter. Also draw the counter's timing diagram. Determine the modulus for the counter.

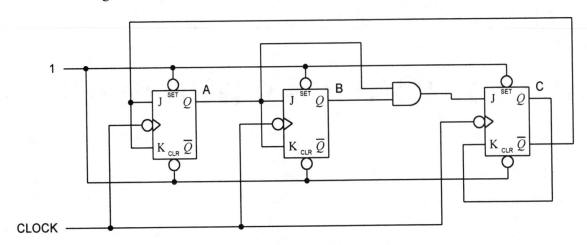

Fig. 11-1 Synchronous counter schematic for example 11-1

To analyze the counter, the circuit excitation (or present state - next state) table given in Table 11-1 is produced. The counter is assumed to start at state 000. The analysis indicates that the counter is a mod-5 counter.

CLOCK	Present State C B A	J_C	K_C	J_B	K_B	J_A	K_A	Next State C B A
0	0 0 0	0	0	0	0	1	1	0 0 1
1	0 0 1	0	0	1	1	1	1	0 1 0
2	0 1 0	0	0	0	0	1	1	0 1 1
3	0 1 1	1	0	1	1	1	1	1 0 0
4	1 0 0	0	1	0	0	0	0	0 0 0
	1 0 1	0	1	1	1	0	0	0 1 1
	1 1 0	0	1	0	0	0	0	0 1 0
	1 1 1	1	1	1	1	0	0	0 0 1

Table 11-1 Complete present state - next state table for example 11-1

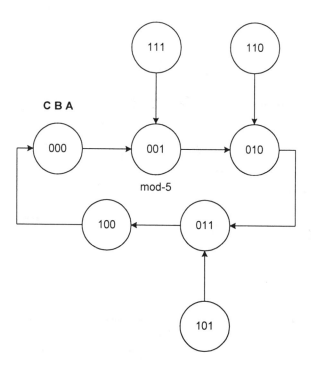

Fig. 11-2 State transition diagram for synchronous counter in example 11-1

Laboratory Projects

Analyze the following synchronous counters. For each counter, determine the counter's modulus and sketch its timing diagram (show the clock input and the flip-flops' Q outputs). Construct and test the operation of the counter circuits. Use an oscilloscope to compare the output waveforms for each of the counters with your theoretical prediction.

11.1   Synchronous counter 1

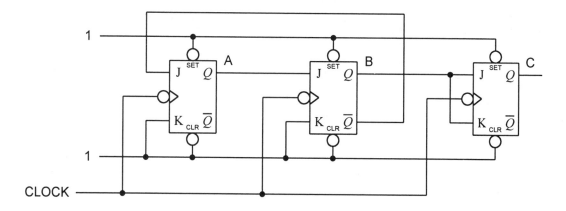

11.2 Synchronous counter 2

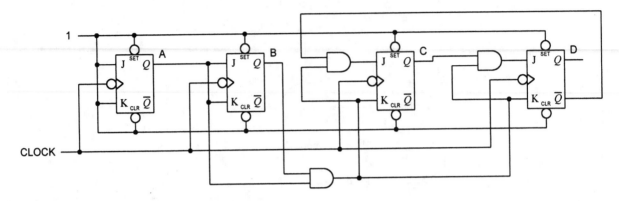

11.3 Synchronous counter 3

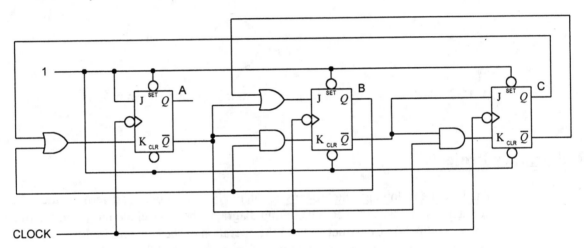

11.4 Synchronous counter 4

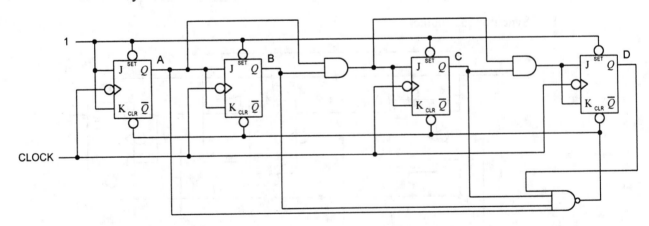

IC COUNTER APPLICATIONS

Objectives

- To control the operation of IC counter chips.
- To produce specified count sequences using IC counter chips.
- To apply IC counter chips in design applications requiring the cascading of chips together.

Suggested Parts					
74LS00	74LS04	74LS08	74LS10	74LS20	74LS47
74LS90	74LS93	74LS160A	74LS190	74LS191	74LS393
MAN72					

IC Counters

Various asynchronous and synchronous counter designs have been integrated into chips to make circuit applications more convenient for the logic designer. Some variations include the number of flip-flops contained in the chip, the counter modulus, synchronous or asynchronous flip-flop triggering, synchronous or asynchronous counter resetting, synchronous or asynchronous counter loading, up/down count control, and various counter cascading implementations. Different feature combinations are found in different chip part numbers. Mod-16 and mod-10 IC counters are commonly available. A mod-10 counter is also referred to as a decade or BCD counter. One common application for counters is in frequency division, in which the input signal (applied to the clock) frequency is divided by a specified factor to produce the resultant output frequency of the divider.

Example 12-1

Design a frequency divider circuit using 74LS90 decade counter chips that will output two different frequencies: 1 and 0.1 MHz. A 5 MHz clock signal is available.

A block diagram for the solution is shown in Fig. 12-1. The resulting schematic is given in Fig. 12-2.

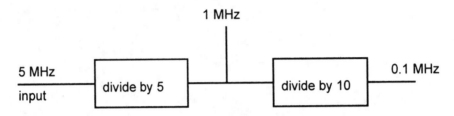

Fig. 12-1 Block diagram for solution to example 12-1

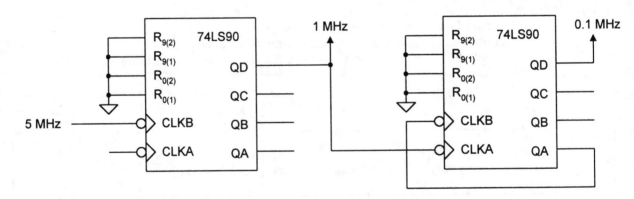

Fig. 12-2 Design schematic for example 12-1

Laboratory Projects

12.1 Specified mod-7 count sequences with a 74LS190
Design mod-7 counters using a 74LS190 that produce the count sequences specified in the following state transition diagrams. Note: Transient states (which <u>may</u> be present) are not shown in the diagrams. Construct and test each counter circuit.

Circuit 1:

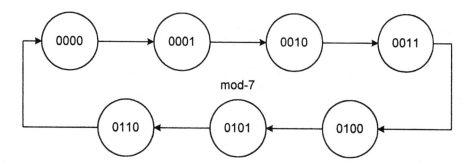

Circuit 2:

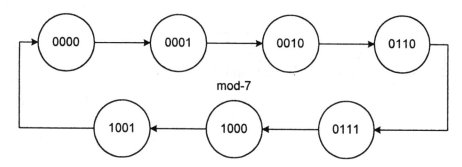

Circuit 3:

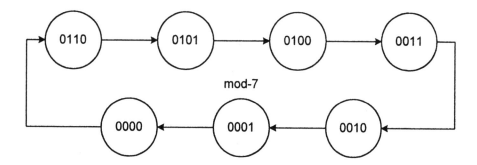

12.2 Specified mod-7 count sequences with a 74LS160ADesign mod-7 counters using a 74LS160A that produce the count sequences specified in the following state transition diagrams. Note: Transient states (which <u>may</u> be present) are not shown in the diagrams. Construct and test each counter circuit.

Circuit 1:

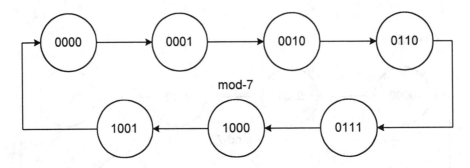

Circuit 2:

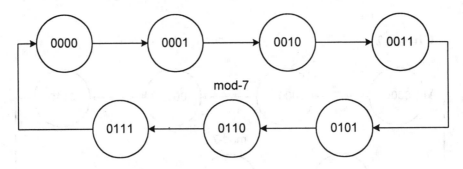

Circuit 3:

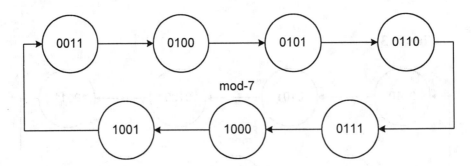

12.3   Mod-100 BCD counter and display
Design a mod-100 counter by cascading the 74LS190 and 74LS160A decade counters together. Display the count sequence on a 2-digit decimal display using two common-anode, 7-segment displays and 74LS47 BCD-to-7-segment decoder drivers. Construct and test the counter and display circuit.

12.4   Frequency divider circuits
Design and test 6 different frequency divider circuits using a 74LS90 decade counter. Obtain each of the following frequency divisions. Use an oscilloscope to monitor the input and output signals of each divider circuit. Sketch the input and output waveforms.

<u>Divide by:</u>
        2 (with a 50% duty cycle)
        5 (with an unspecified duty cycle)
        10 (with a 20% duty cycle)
        10 (with a 40% duty cycle)
        10 (with a 50% duty cycle)
        9 (with an unspecified duty cycle)

12.5   Divide-by-60 frequency divider
Design 3 different logic circuits using a 74LS393 dual 4-bit binary counter chip that will divide the frequency of an input square wave signal by a factor of 60. Use a frequency counter to verify proper operation of your circuit designs. Construct and test each counter circuit.

12.6   Frequency divider
Design a frequency divider circuit that will output four pulse frequencies: 2.5 KHz, 1.0 KHz, 625 Hz, and 500 Hz. Assume that only a 5.0 KHz TTL signal is available for the divider input. Use only two chips, one 74LS90 and one 74LS93. Construct and test the frequency divider circuit. Use a frequency counter to verify proper operation of your circuit design.

12.7   Programmable frequency divider

Design a programmable frequency divider circuit using a 74LS191 up/down binary counter chip. The frequency dividing factor (F) will be a 4-bit binary number $(1 \le F \le 15)$, which will be input from logic switches. Hint: Count down from F and asynchronously re-load F when the counter reaches 0000. Construct and test the frequency divider circuit. Use a frequency counter to verify proper operation of your circuit design.

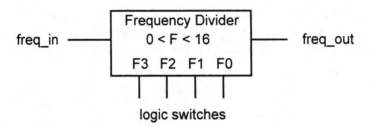

12.8   Self-stopping decade counter

Design a self-stopping decade counter using a 74LS190. The counter should display the count sequence of 9 down to 0 on a 7-segment display. The count sequence will be started with an active-high signal called START. The count will stop at zero. Construct and test the counter and display circuit.

12.9   Divide-by-120 frequency divider

Design a frequency divider circuit that will divide the input signal frequency by 120. Use a 74LS190 and a 74LS191 counter chip. Construct and test the frequency divider circuit. Use a frequency counter to verify proper operation of your circuit design.

12.10   Frequency divider

Design a frequency divider circuit that will output three pulse frequencies: 25 KHz, 10 KHz, and 2 KHz. Assume that only a 150 KHz TTL signal is available for the divider input. Use any available IC counter chips. Construct and test the frequency divider circuit. Use a frequency counter to verify proper operation of your circuit design.

12.11   Mod-36 BCD counter

Design a mod-36 BCD counter by cascading the 74LS190 and 74LS160A decade counters together. The recycling count sequence should be from 00 to 35 in 8421 BCD.

SEQUENTIAL CIRCUIT DESIGN USING PROGRAMMABLE LOGIC DEVICES

Objective

- To design and implement sequential logic circuits using programmable logic devices (PLDs).

Suggested Part
GAL16V8

Sequential Circuits Using PLDs

Many programmable logic devices contain flip-flops and, therefore, can also be used to implement sequential circuits. The design and implementation of sequential circuits using CUPL is carried out in much the same fashion as for combinatorial circuits. The flip-flops contained in most PLDs are D type flip-flops in which the D inputs are produced by the programmable AND/OR gate structure in the PLD. A registered output pin is defined in the logic equations by using the pin modifier ".D" attached to the output pin name. CUPL also provides a convenient means to define many types of sequential circuits called state machine entry. State machine entry uses a "present-state, next-state" format to define the desired sequence of states. The operation of a sequential circuit can also be easily controlled by external signals in the state machine definition. While very flexible and easy to implement for relatively low modulus counters, the state machine entry is somewhat cumbersome for higher mods. The high level language framework of CUPL provides many other options for defining sequential circuits.

The GAL16V8 can be programmed to provide registered outputs on any of its eight output pins. Whenever at least one of the outputs from the GAL16V8 uses an internal D flip-flop to produce the output, the chip is said to be programmed in the registered mode. **In the registered mode of operation, pin 1 is automatically configured to be the clock pin for all flip-flops used (synchronous clocking) and pin 11 is the common tristate output enable pin (active-low) for all registered outputs (but not combinational outputs).** Neither of these two pins can be defined to be any other type of input to the PLD when it is configured for a registered output.

Example 13-1

Design a counter using a GAL16V8 that produces the 4-bit, irregular, recycling count sequence given in the timing diagram of Fig. 13-1.

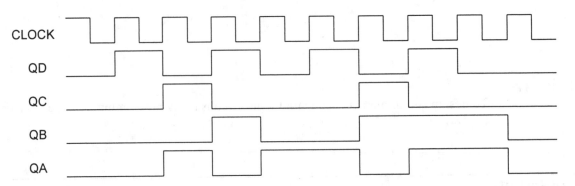

Fig. 13-1 Timing diagram for sequential design example 13-1

A CUPL logic description file solution is shown in Fig. 13-2. Pin 1 is the clock input and pin 11 is the tristate output enable control for the flip-flops. The tristate control was not specified for this application and, therefore, pin 11 should be grounded to permanently enable the flip-flop outputs for this counter. The "$define" command is one of CUPL's preprocessor commands. The "$define" command is used to replace a given character string by another specified operator, number, or variable name. In this example file, each of the 16 binary combinations that is possible with 4-bits is given the variable names of S0 through S15. The names are arbitrary and this application will use only nine of the states, but all of the possible 4-bit counter states have been defined for our convenience so that any 4-bit counter sequence can be easily defined by minor changes in this file. **The "$" symbol is the first character for all preprocessor commands and it must be placed in column 1 of the line. Preprocessor command lines do not end with a semicolon.** The logic definition (equations) section uses the state machine entry format (indicated by the keyword "sequence") to describe the sequence for this counter. The output bits of the counter are listed after "sequence."

```
Name       IRREGSEQ;            Partno    L155-17A;
Date       07/17/93;            Revision  01;
Designer   Greg Moss;           Company   Digi-Lab;
Assembly   Example board;       Location  U121;
Device     G16V8A;              Format    j;

/*************************************************************/
/*   Synchronous counter with irregular sequence         */
/*    0 -> 8 -> 5 -> A -> 1 -> 9 -> 6 -> B -> 3 -> 0    */
/*************************************************************/

/**   Inputs   **/
Pin  1  =  CLK;                       /* CLOCK INPUT    */
Pin 11  =  !OE;                       /* OUTPUT ENABLE  */

/**   Outputs   **/
Pin [14..17]  =  [QA,QB,QC,QD];    /* COUNTER OUTPUTS */

/** Declarations and Intermediate Variable Definitions **/
field irregcnt = [QD,QC,QB,QA];          /* counter bits */

$define S0   'b'0000              /* define counter states */
$define S1   'b'0001
$define S2   'b'0010
$define S3   'b'0011
$define S4   'b'0100
$define S5   'b'0101
$define S6   'b'0110
$define S7   'b'0111
$define S8   'b'1000
$define S9   'b'1001
$define S10  'b'1010
$define S11  'b'1011
$define S12  'b'1100
$define S13  'b'1101
$define S14  'b'1110
$define S15  'b'1111

/** Logic Definition **/
sequence    irregcnt    {  present S0     next S8;
                           present S8     next S5;
                           present S5     next S10;
                           present S10    next S1;
                           present S1     next S9;
                           present S9     next S6;
                           present S6     next S11;
                           present S11    next S3;
                           present S3     next S0;
                        }
```

Fig. 13-2 CUPL design logic description file for example 13-1

In this example, the output bits have been assigned the field name "irregcnt". The braces {} mark the beginning and end of the state machine description. The sequence desired is represented in a series of present state and next state combinations. The desired sequence can be easily changed by simply changing the order of the states. Each present state must be unique in the sequence description.

Example 13-2

Design an up/down, recycling decade counter using a GAL16V8. The mod-10 counter also has count enable, count direction, and synchronous counter reset controls. The counter should also produce a ripple carry output signal that goes low for the last state in the up count or down count sequence.

CLR	ENABLE	DIR	function
0	0	X	HOLD
0	1	0	COUNT UP
0	1	1	COUNT DOWN
1	X	X	CLEAR

A CUPL design solution is shown in Fig. 13-3. In this logic description file, the count enable and direction control bits are grouped together as a field and intermediate variables are defined for the count up and count down modes of operation. The ten counter states are given labels with the "$define" command and the count sequence is described in the state machine format. In this case, the next state that will be produced by the counter is dependent on the three external control inputs (CLR, ENABLE, and DIR). For each present state, if the counter is not being cleared and the "up" bit combination is asserted, then the next state will be the resulting up count state. If, on the other hand, the counter is not being cleared and the "down" bit combination is asserted, then the resulting next state would be the previous value in a down count sequence. If CLR is active, then the next state will always be 0000. The default condition would exist if the counter is not enabled and the count should remain in the same state. The ripple carry output signal should be asserted under two different circumstances. It should be active during the state "S0" if the counter is counting down or during state "S9" if the counter is counting up. The keyword "OUT" is used to assert the combinational output signal "ripple" (which is active-low) during the proper state.

```
Name       MOD10;                    Partno    L155-17A;
Date       07/17/93;                 Revision  01;
Designer   Greg Moss;                Company   Digi-Lab;
Assembly   Example board;            Location  U129;
Device     G16V8A;                   Format    j;

/*******************************************************/
/* up/down decade counter with synchronous clear       */
/* counter also has an asynchronous ripple carry output */
/*******************************************************/

/**   Inputs   **/
pin 1 = CLK;             /* Counter clock               */
pin 2 = CLR;             /* Counter clear input         */
pin 3 = DIR;             /* Counter direction input     */
pin 4 = ENABLE;          /* Counter enable              */
pin 11 = !OE;            /* Register output enable       */

/**   Outputs   **/
pin [19..16] = [Q3..0];         /* Counter outputs      */
pin 15 = !ripple;               /* Ripple carry out     */

/** Declarations and Intermediate Variable Definitions **/
field counter = [Q3..0];        /* counter bit field  */
field mode = [ENABLE,DIR];      /* mode control field */

up = mode:2;                    /* define count up mode   */
down = mode:3;                  /* define count down mode */

$define S0 'b'0000              /* define counter states  */
$define S1 'b'0001
$define S2 'b'0010
$define S3 'b'0011
$define S4 'b'0100
$define S5 'b'0101
$define S6 'b'0110
$define S7 'b'0111
$define S8 'b'1000
$define S9 'b'1001

/** Logic Equations **/

sequence counter     {
    present S0   if  !CLR  &  up        next S1;
                 if  !CLR  &  down      next S9;
                 if   CLR               next S0;
                 default                next S0;
                      if down    OUT    ripple;
    present S1   if  !CLR  &  up        next S2;
                 if  !CLR  &  down      next S0;
                 if   CLR               next S0;
                 default                next S1;
    present S2   if  !CLR  &  up        next S3;
                 if  !CLR  &  down      next S1;
                 if   CLR               next S0;
                 default                next S2;
```

97

```
present S3      if  !CLR  &  up        next S4;
                if  !CLR  &  down      next S2;
                if   CLR               next S0;
                default                next S3;
present S4      if  !CLR  &  up        next S5;
                if  !CLR  &  down      next S3;
                if   CLR               next S0;
                default                next S4;
present S5      if  !CLR  &  up        next S6;
                if  !CLR  &  down      next S4;
                if   CLR               next S0;
                default                next S5;
present S6      if  !CLR  &  up        next S7;
                if  !CLR  &  down      next S5;
                if   CLR               next S0;
                default                next S6;
present S7      if  !CLR  &  up        next S8;
                if  !CLR  &  down      next S6;
                if   CLR               next S0;
                default                next S7;
present S8      if  !CLR  &  up        next S9;
                if  !CLR  &  down      next S7;
                if   CLR               next S0;
                default                next S8;
present S9      if  !CLR  &  up        next S0;
                if  !CLR  &  down      next S8;
                if   CLR               next S0;
                default                next S9;
                    if up       OUT  ripple;
                }
```

Fig. 13-3 CUPL design logic description file for example 13-2

Example 13-3

Design a mod-100, recycling, BCD counter using a GAL16V8. The two-digit decade counter also needs to have an active-low count enable control EN.

A CUPL design solution is shown in Fig. 13-4. Since describing the count sequence for a mod-100 counter in state machine format would be rather long and tedious, logic equations have been written for the two decade counters. The 4-bits for each of the two decade counters have been grouped together as a field. In writing each of the two decade counter equations, the extension ".D" is used to indicate that the outputs are to be registered outputs. The GAL16V8 will be in registered mode and the equations represent the inputs to each of the D flip-flops. The logic equation for the LSD counter can be interpreted as: If the counter is enabled, after being clocked the LSD will be equal to the decimal value 1 if the LSD currently is equal to 0, or it will be equal to a decimal 2 if it is currently equal to 1, or to 3 if it is currently 2, etc. If the counter is not enabled, then the LSD counter should remain in the same state after it is clocked.

```
Name        MOD-100;          Partno    L155-17C;
Date        07/17/93;         Revision  01;
Designer    Greg Moss;        Company   Digi-Lab;
Assembly    Example board;    Location  U126;
Device      G16V8A;           Format    j;

/*******************************************************/
/* Synchronous MOD-100 BCD up counter with enable     */
/*******************************************************/

/**   Inputs  **/
Pin  1  =  CLK;                      /* Clock input        */
Pin  2  =  !EN;                      /* Count enable       */
Pin 11  =  !OE;            /* Register output enable    */

/**   Outputs  **/
Pin [12..15] = [L0..3];       /* Counter LSD output   */
Pin [16..19] = [M0..3];       /* Counter MSD output   */


/**   Declarations and Intermediate Variable Definitions
**/

field LSD = [L3..0];
field MSD = [M3..0];


/**   Logic Equations  **/

LSD.D  =  (   ['d'1]   &   LSD:0
          #   ['d'2]   &   LSD:1
          #   ['d'3]   &   LSD:2
          #   ['d'4]   &   LSD:3
          #   ['d'5]   &   LSD:4
          #   ['d'6]   &   LSD:5
          #   ['d'7]   &   LSD:6
          #   ['d'8]   &   LSD:7
          #   ['d'9]   &   LSD:8
          #   ['d'0]   &   LSD:9  )  &   EN
          #   LSD                   &   !EN;

MSD.D  =  ((  ['d'1]   &   MSD:0
          #   ['d'2]   &   MSD:1
          #   ['d'3]   &   MSD:2
          #   ['d'4]   &   MSD:3
          #   ['d'5]   &   MSD:4
          #   ['d'6]   &   MSD:5
          #   ['d'7]   &   MSD:6
          #   ['d'8]   &   MSD:7
          #   ['d'9]   &   MSD:8
          #   ['d'0]   &   MSD:9  )  &  LSD:9
          #   MSD   &   LSD:[0..8]  )  &   EN
          #   MSD                     &   !EN;
```

Fig. 13-4 CUPL design logic description file for example 13-3

The MSD equation is similar, but it must also take into account the state of the LSD counter to determine its count action. The least significant digit must control the most significant digit to count 0 through 9, and then 10 through 19, and 20 through 29, etc. The MSD will increment on the next clock only when the LSD is equal to 9. The MSD should remain in the same state while LSD is in the states of 0 through 8.

Example 13-4

Design a mod-64, recycling, up counter using a GAL16V8. The counter will also be used to generate two active-high timing signals (TIME1 and TIME2). TIME1 should be asserted when the 6-bit counter is in the hexadecimal states 16 through 20 (inclusive). TIME2 should be asserted when the counter is in the hexadecimal states 2A through 3C.

A CUPL design solution is shown in Fig. 13-5. The logic equations have the ".d" extension to produce registered outputs in the GAL16V8. The least significant bit must change states with each clock cycle. Therefore, the inverse of the current output is fed into the D input of flip-flop Q0. The XOR function can be used to determine the input for each of the other D-type flip-flops. In the logic equations for Q2 through Q5, the D input for each flip-flop will be the current output bit exclusive ORed with the ANDing of the set of bits that are lower in significance than this output bit.

A convenient shortcut notation is utilized in this solution where a set of variables given in the brackets are to be operated on with the function specified after the equality operator. For example:

```
[Q1..0]:&  is equivalent to  (Q1 & Q0)
[Q2..0]:&  is equivalent to  (Q2 & Q1 & Q0)
[Q3..0]:&  is equivalent to  (Q3 & Q2 & Q1 & Q0)
etc.
```

```
Name        MOD-64;              Partno     L155-17D;
Date        07/17/93;            Revision   01;
Designer    Greg Moss;           Company    Digi-Lab;
Assembly    Example board;       Location   U142;
Device      G16V8A;              Format     j;

/***********************************************************/
/* mod-64 counter defined using the XOR function and the   */
/* ANDing of a set of variables with the equality operator */
/***********************************************************/

/** Inputs **/

pin 1 = CLK;                /* Counter clock input     */
pin 11 = !OE;               /* Register output enable  */

/** Outputs **/

pin [19..14] = [Q5..0];     /* Counter outputs         */
pin 13 = TIME1;             /* Output waveform #1       */
pin 12 = TIME2;             /* Output waveform #2       */


/** Declarations and Intermediate Variable Definitions  **/

field  counter = [Q5..0];


/** Logic Equations **/

Q0.d   =     !Q0;
Q1.d   =     (Q1  $  Q0);
Q2.d   =     (Q2  $  [Q1..0]:&);
Q3.d   =     (Q3  $  [Q2..0]:&);
Q4.d   =     (Q4  $  [Q3..0]:&);
Q5.d   =     (Q5  $  [Q4..0]:&);

TIME1  =     counter:[16..20];
TIME2  =     counter:[2A..3C];
```

Fig. 13-5 CUPL design logic description file for example 13-4

Example 13-5

Design a mod-16, recycling, up/down, binary counter using a GAL16V8. The counter can also be synchronously parallel loaded with a 4-bit number (D3 D2 D1 D0) or synchronously reset to 0. The 3-bit function control (M2 M1 M0) for the counter is described in the following table.

M2	M1	M0	Function
0	0	X	HOLD
0	1	X	LOAD
1	0	0	COUNT UP
1	0	1	COUNT DOWN
1	1	X	RESET

A CUPL solution file is shown in Fig. 13-6. In this design, bit fields are declared and all 4-bit combinations are assigned names to use. The various functional modes are also given convenient variable names. The complex equation describes the logic for each of the five different functional modes of operation for the counter.

Example 13-6

Design a mod-16, recycling, binary counter using a GAL16V8. The counter should have an active-low enable and a synchronous active-high reset.

An example CUPL solution file using the "$repeat" preprocessor command is shown in Fig. 13-7. This preprocessor command allows a set of CUPL statements to be repeated a number of times, controlled by an index variable. The preprocessor command "$repend" identifies the end of the set of statements that are to be repeated. **The "$repend" statement must be on a line by itself, with no spaces following the statement.** Also remember that all preprocessor commands start with the "$" symbol in column 1 of the line.

In this design, the index variable is called i and ranges in value from 0 to 15. The index variable determines the number of times that the "present - next" statements should be repeated in the loop. The index "i" will be equal to 0 the first time through the loop, then 1 the next time, etc., until it is finally equal to 15, after which, the loop will be terminated. The current index value is used in determining the present state and next state each time through the loop. An arithmetic operation (adding 1 to the current index value to count up) is used for the next state. The counter's modulus is set by the percent % symbol in the next statement. If the %(16) were changed to %(12), for example, a mod-12 counter would be created instead and states s12 through s15 would feed back into earlier states in the count sequence.

```
Name        LOADCNTR;              Partno    L155-17E;
Date        07/20/93;              Revision  01;
Designer    G. Moss;               Company   Digi-Lab;
Assembly    Controller;            Location  U922;
Device      G16V8A;                Format    j;

/******************************************************/
/*  Mod-16 binary up/down counter with parallel load  */
/******************************************************/

              /****  Inputs  ****/

pin 1 = clk;              /* Counter clock            */
pin [2..4] = [M2..0];     /* Counter function controls */
pin [5..8] = [D3..0];     /* Parallel data inputs      */
pin 11 = !oe;             /* Register output enable    */


              /****  Outputs  ****/

pin [12..15] = [Q0..3];     /* Counter outputs         */


              /**** Definitions ****/

field counter = [Q3..0];    /* counter bit field       */
field data    = [D3..0];    /* data input field        */
field mode    = [M2..0];    /* function control field  */

hold      =   mode:[0..1];  /* define functions        */
load      =   mode:[2..3];
count_up  =   mode:4;
count_dn  =   mode:5;
reset     =   mode:[6..7];

$define S0   'b'0000        /* define counter states   */
$define S1   'b'0001
$define S2   'b'0010
$define S3   'b'0011
$define S4   'b'0100
$define S5   'b'0101
$define S6   'b'0110
$define S7   'b'0111
$define S8   'b'1000
$define S9   'b'1001
$define S10  'b'1010
$define S11  'b'1011
$define S12  'b'1100
$define S13  'b'1101
$define S14  'b'1110
$define S15  'b'1111
```

103

```
                /**** Logic Equations ****/
counter.d    =
                                    /* hold data mode      */
         counter  &  hold
                                    /* parallel load mode */
    #   data      &  load
                                    /* count up mode       */
    #   (    S1   &   counter:S0
        #  S2   &   counter:S1
        #  S3   &   counter:S2
        #  S4   &   counter:S3
        #  S5   &   counter:S4
        #  S6   &   counter:S5
        #  S7   &   counter:S6
        #  S8   &   counter:S7
        #  S9   &   counter:S8
        #  S10  &   counter:S9
        #  S11  &   counter:S10
        #  S12  &   counter:S11
        #  S13  &   counter:S12
        #  S14  &   counter:S13
        #  S15  &   counter:S14
        #  S0   &   counter:S15 )  &   count_up
                                    /* count down mode      */
    #   (    S15  &   counter:S0
        #  S0   &   counter:S1
        #  S1   &   counter:S2
        #  S2   &   counter:S3
        #  S3   &   counter:S4
        #  S4   &   counter:S5
        #  S5   &   counter:S6
        #  S6   &   counter:S7
        #  S7   &   counter:S8
        #  S8   &   counter:S9
        #  S9   &   counter:S10
        #  S10  &   counter:S11
        #  S11  &   counter:S12
        #  S12  &   counter:S13
        #  S13  &   counter:S14
        #  S14  &   counter:S15 )  &   count_dn
                                    /* reset mode          */
    #   S0   &   reset;
```

Fig. 13-6 CUPL design logic description file for example 13-5

```
Name       counter;           Partno    L159-11;
Date       12/08/96;          Rev       01;
Designer   Greg Moss;         Company   Digi-Lab, Inc.;
Assembly   Counter Board;     Location  U333;
Device     G16V8A;            Format    j;

/**************************************************/
/* mod-16 counter with active-low enable          */
/**************************************************/

/** Inputs **/
Pin 1 = CLK;
Pin 2 = !EN;                    /* active-low enable  */
Pin 3 = RESET;
Pin 11 = !OE;

/** Outputs **/
Pin [12..15] = [Q0..3];    /*  outputs              */

/** Declarations & Intermediate Variable Definitions **/
field COUNT = [Q3..0];      /* COUNTER                 */

$define s0   'b'0000
$define s1   'b'0001
$define s2   'b'0010
$define s3   'b'0011
$define s4   'b'0100
$define s5   'b'0101
$define s6   'b'0110
$define s7   'b'0111
$define s8   'b'1000
$define s9   'b'1001
$define s10  'b'1010
$define s11  'b'1011
$define s12  'b'1100
$define s13  'b'1101
$define s14  'b'1110
$define s15  'b'1111

/** Equations **/
sequence COUNT
{
$repeat i=[0..15]
  present s{i}    if EN & !RESET   next s{(i+1)%(16)};
                  if RESET         next s{0};
                  default          next s{i};
$repend
}
```

Fig. 13-7 CUPL design logic description file for example 13-6

Laboratory Projects

13.1 Custom sequential circuit
Design a custom sequential circuit using a GAL16V8 that will produce the sequence given in the following state diagram.

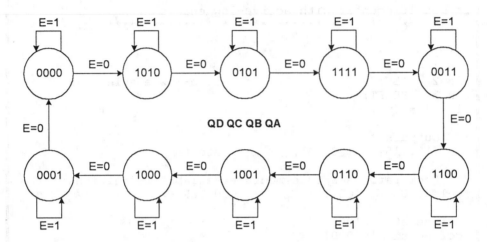

13.2 Gray code counter
Design a 4-bit, up/down, recycling Gray code counter using a GAL16V8. The count direction is controlled by a signal called F, as indicated in the following function table. Label counter outputs Q_3, Q_2, Q_1, and Q_0. The circuit also should produce an output signal called INDEX, which goes low whenever the counter state is 0000.

F	operation
0	Count Up
1	Count Down

Gray code sequence

	Q3	Q2	Q1	Q0	
	0	0	0	0	
count	0	0	0	1	⇑
up	0	0	1	1	count
sequence	0	0	1	0	down
⇓	0	1	1	0	sequence
	0	1	1	1	
	0	1	0	1	
	0	1	0	0	
	1	1	0	0	
	1	1	0	1	
	1	1	1	1	
	1	1	1	0	
	1	0	1	0	
	1	0	1	1	
	1	0	0	1	
	1	0	0	0	

13.3 Mod-16 binary counter
Design a mod-16, recycling, binary counter using a GAL16V8. The counter's function table is given below. The counter should also produce an active-low ripple carry output signal called CARRY.

C1	C0	operation
0	0	Reset
0	1	Count Down
1	0	Count Up
1	1	Hold Count

13.4 Mod-60 BCD counter
Design a mod-60, recycling, BCD counter using a GAL16V8. The counter should have an active-low reset (RES) signal . The count sequence will be 0 through 59_{10} in BCD.

13.5 Mod-128 binary counter
Design a mod-128, recycling, binary up-counter using a GAL16V8. The counter should have an active-high count enable called EN. Also output a single active-high pulse (called PULSE) during each mod-128 count sequence. The pulse will end when the counter recycles. The pulse width will be controlled by an input signal called W as shown in the table below.

W	PULSE width
0	4 clock periods
1	8 clock periods

13.6 Mod-200 binary counter
Design a mod-200, recycling, binary up-counter using a GAL16V8. Hint: Define an intermediate variable that decodes the desired final counter state and then include this variable in the logic equations for an 8-bit binary counter.

13.7 Stepper motor sequence controller

Design a half-step sequencer to control a stepper motor using a GAL16V8. The sequencer should produce the appropriate sequence of states to drive the motor either clockwise (CW) or counterclockwise (CCW). The stepper direction is controlled by the signal CW. The stepper enable is called STEP. The function table is given below. Make sure the sequencer is self-starting.

STEP	CW	function
0	X	HALT
1	0	CCW
1	1	CW

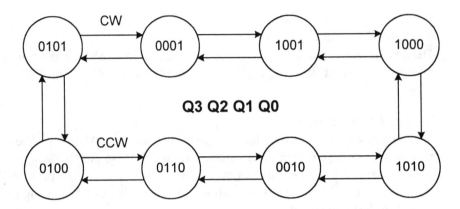

Half-step sequence for stepper motor control

13.8 Variable frequency divider

Design a variable frequency divider using a GAL16V8. The frequency divider should divide the input frequency by one of four different factors. The divide-by factor is controlled by two mode controls as described by the following function table. The mode controls are used to change the modulus of the counter used for the frequency division. The output waveform will be high for two clock cycles starting at state zero.

M1	M0	divide by:
0	0	5
0	1	10
1	0	12
1	1	15

FREQ_IN $\rangle$ variable frequency divider $\rangle$ FREQ_OUT

input frequency   output frequency

M1   M0

13.9   Digital lock
Design a digital lock circuit using a GAL16V8. The lock will have a 4-bit data input
and an ENTER signal. Use a mod-5 binary counter to keep track of the sequencing
through the combination. The ENTER signal is the clock signal for the counter. A 4-
number lock combination sequence (established in the design) will be required to
unlock the combination lock. Each input number (4-bit value) will be applied and then
the ENTER signal will be asserted (active-high). The machine states must be
sequenced in the proper order to unlock the lock. The "START" state, or STATE0, is
followed by the intermediate states STATE1, then STATE2, then STATE3, and then
finally STATE4, which "UNLOCKs" the lock. If an incorrect input combination is
"ENTERed" during any counter state, the counter will return to STATE0. The state
transition diagram for the digital lock is shown below.

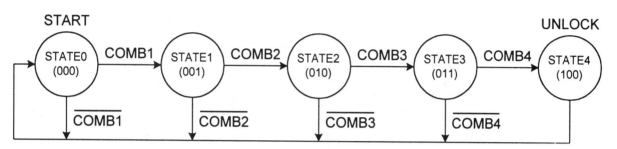

Returns to STATE0 if any incorrect combination is applied

13.10   Mod-24 binary counter
Design a mod-24, recycling, binary counter using a GAL16V8. The counter should
have two enables, an active-low enable EN0 and an active-high enable EN1. The
counter also should have a synchronous active-low reset named RESET.

SHIFT REGISTER APPLICATIONS

Objectives

- To apply shift registers in serial and parallel digital data transfer applications.
- To apply shift registers in counter applications.
- To design and implement a shift register application in a PLD.

Suggested Parts					
74LS75	74LS95B	74LS160A	74LS164	74166A	74LS190
74LS393	GAL16V8				

Shift Registers

Registers consist of a set of flip-flops used to store and transfer binary data in a digital system. Registers can be classified according to the types of input and output data movement. With the two basic forms of data transfer, serial and parallel, there are the following categories of registers:

1. Parallel-in/Parallel-out (PIPO)
2. Serial-in/Serial-out (SISO)
3. Parallel-in/Serial-out (PISO)
4. Serial-in/Parallel-out (SIPO)

Many MSI shift register chips are designed to handle data movement into or out of the register in any desired manner. The input data may be either serial or parallel and the output data may be serial or parallel. The 74LS95 (shown in Fig. 14-1) is an MSI, 4-bit shift register with this kind of flexibility. The serial data input is labeled SER and the parallel data inputs are D, C, B, and A. The MODE control determines whether serial or parallel data flow will be used. Since all four register outputs (Q_D, Q_C, Q_B, and Q_A) are available, the data output may be either serial or parallel. For added flexibility, this chip also has two separate clock inputs: one for serial mode and another for parallel mode of operation.

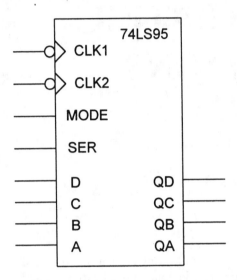

Fig. 14-1 74LS95 4-bit shift register with parallel access

Example 14-1

Design an 8-bit universal shift register using a GAL20V8. Note that a GAL20V8 is specified here so that there will be enough input and output pins available for this application. The 8-bit shift register can move data serially or in parallel. The serial data movement can be either shifting the data to the left (from Q7 towards Q0) or to the right (from Q0 toward Q7). A single serial data input (SER_IN) is used during serial shifting in either direction. The parallel data inputs are labeled D7 through D0. Register function is controlled by S1 and S0 as described in the following function table.

S1	S0	Function
0	0	HOLD
0	1	LOAD
1	0	SHIFT LEFT
1	1	SHIFT RIGHT

An example CUPL design logic description file is shown in Fig. 14-2. Bit fields have been defined for the set of parallel inputs, register outputs, and the two register function control bits. The combination of bits that represent data shifting one bit position to the left and to the right are also defined as bit fields. The location of the serial input bit (SER_IN) is dependent on the direction of the serial shifting operation. Variable names for the four functions of the shift register are also defined. The logic equation for the register "output" is equal to the appropriate bit field combination for each of the function names.

Example 14-2

Design an 8-bit, registered, barrel shifter using a GAL20V8A. Note that a GAL20V8A is specified here so that there will be enough input and output pins available for this application. The 8-bit barrel shifter has 8 data inputs (D7 through D0) that can be cyclically rotated from 0 to 7 places under the control of the select inputs (S2, S1, and S0). The resultant output data is to be stored in a register.

An example CUPL design logic description file is shown in Fig. 14-3. Bit fields have been declared for the shift register "output" and for the input that controls the number of bit places for the data to be rotated. The logic equation for "output" then matches the list of the appropriate data bit positions with the number of requested data shifts.

```
Name       UNIV_S-R;              Partno    L155-18A;
Date       07/21/93;              Revision  01;
Designer   G. MOSS;               Company   Digi-Lab;
Assembly   I/O board;             Location  U823;
Device     G20V8A;                Format    j;

          /**********************************************/
          /*  8-bit universal shift register          */
          /*    S1  S0    Function                     */
          /*     0   0    HOLD                         */
          /*     0   1    LOAD                         */
          /*     1   0    SHIFT LEFT                   */
          /*     1   1    SHIFT RIGHT                  */
          /**********************************************/
          /*  Note Target Device:  GAL20V8A           */
          /**********************************************/


/**   Inputs   **/
Pin   1    =   CLK       ;   /* Clock input                */
Pin  [2,3]  =  [S1,S0]   ;   /* Register controls          */
Pin  [4..11] = [D7..D0]  ;   /* Data inputs (parallel load) */
Pin  14    =   SER_IN    ;   /* Serial input data          */
Pin  13    =   !OE       ;   /* Output enable              */


/**   Outputs   **/
Pin [15..22] = [Q0..Q7]  ;   /* Register outputs           */


/** Declarations and Intermediate Variable Definitions **/

field datain    =   [D7..D0]           ;
field output    =   [Q7..Q0]           ;
field leftshft  =   [Q6..Q0,SER_IN]    ;
field rghtshft  =   [SER_IN,Q7..Q1]    ;
field contrl    =   [S1,S0]            ;

HOLD  =   contrl:0 ;
LOAD  =   contrl:1 ;
LEFT  =   contrl:2 ;
RGHT  =   contrl:3 ;


/**  Logic Equations  **/

output.d  =
            output    &   HOLD
        #   datain    &   LOAD
        #   leftshft  &   LEFT
        #   rghtshft  &   RGHT  ;
```

Fig. 14-2 CUPL logic description file for example 14-1

```
Name        Barrel;              Partno    L155-18B;
Date        07/21/93;            Revision  01;
Designer    G. Moss;             Company   Digi-Lab;
Assembly    I/O board;           Location  U807;
Device      G20V8A;              Format    j;

/******************************************************/
/* 8-Bit Registered Barrel Shifter                    */
/*    8 data inputs are cyclically rotated from 0 to 7 */
/*    places under control of the select inputs.      */
/******************************************************/
/*    NOTE:  A GAL20V8A is specified as the device    */
/******************************************************/

/**   Inputs   **/
PIN 1            = clock;       /* Register clock       */
PIN [2..9]       = [D7..0];     /* Data inputs          */
PIN [10,11,14]   = [S2..0];     /* Shift select inputs  */
PIN 13           = !oe;         /* Register output enable */

/**   Outputs   **/
PIN [15..22] = [Q7..0];         /* Register outputs     */

/** Declarations and Intermediate Variable Definitions **/

field shift  = [S2..0];         /* Shift select field   */
field output = [Q7..0];         /* Output bits field    */


/** Logic Equations **/

output.d  =    [D7, D6, D5, D4, D3, D2, D1, D0]  &  shift:0
          #    [D0, D7, D6, D5, D4, D3, D2, D1]  &  shift:1
          #    [D1, D0, D7, D6, D5, D4, D3, D2]  &  shift:2
          #    [D2, D1, D0, D7, D6, D5, D4, D3]  &  shift:3
          #    [D3, D2, D1, D0, D7, D6, D5, D4]  &  shift:4
          #    [D4, D3, D2, D1, D0, D7, D6, D5]  &  shift:5
          #    [D5, D4, D3, D2, D1, D0, D7, D6]  &  shift:6
          #    [D6, D5, D4, D3, D2, D1, D0, D7]  &  shift:7;
```

Fig. 14-3 CUPL logic description file for example 14-2

Feedback Shift Register Counters

A count sequence can be generated with a shift register by using the contents of the register to produce a feedback signal for the serial input to the shift register. This type of counter generally requires very little hardware to construct, and the circuitry required to decode the count sequence can be very simple. Ring counters and Johnson counters are examples of feedback shift register counters.

A ring counter has a modulus equal to the number of flip-flops being used in the shift register. There are two variations of the ring counter. One has a single bit that is high and is moved from one flip-flop in the shift register to the next. At the output end of the serial shift register, one is rotated back to the input of the shift register. The other choice is to rotate a single 0 bit throughout the length of the shift register. Decoding of the ring counter is simply done by noting which bit position contains the single 1 (or single 0). Self-starting ring counters can be easily designed by generating a simple feedback signal that monitors all of the shift register bits except the final bit in the chain. For a ring counter that rotates a single 1, use a NOR function. For a ring counter that rotates a single 0, use a NAND function. The fan-in on the appropriate logic gate is always one less than the number of flip-flops in the shift register counter.

The Johnson counter (also called a twisted ring counter) uses the flip-flops a little more efficiently to produce a modulus that is two times the number of flip-flops being used in the shift register. The feedback signal needed for the Johnson sequence is accomplished by merely inverting the single output bit of the serial shift register. Multiple sequences are produced with this simple feedback circuit, but the standard Johnson code sequence includes the state where all flip-flops are low. The Johnson counter sequence can be decoded by monitoring an appropriate pair of bits for each counter state.

Number of flip-flops	Counter modulus	XNOR gate inputs
2	3	Q1, Q0
3	7	Q2, Q1
4	15	Q3, Q2
5	31	Q4, Q2
6	63	Q5, Q4
7	127	Q6, Q5
8	255	Q7, Q5, Q4, Q3
9	511	Q8, Q4
10	1023	Q9, Q6
11	2047	Q10, Q9
12	4095	Q11, Q10, Q9, Q1

Table 14-1 LFSR counter modulus and feedback signals

A pseudo-random count sequence can be generated by a linear feedback shift register (LFSR). The feedback for this simple type of shift register counter is produced by XNORing two or more flip-flop outputs from the shift register. A shift register counter using this feedback arrangement with n flip-flops would have a count modulus of $2^n - 1$. The count modulus for any length LFSR counter can be shortened by causing it to skip an appropriate number of states. Table 14-1 lists the proper inputs to the XNOR feedback gate that is generating the serial input to flip-flop Q0 (the first flip-flop in the shift register chain).

Example 14-3

Design a self-starting, mod-7 ring counter using a GAL16V8. The ring counter should output a single 1 that is rotated through the shift register.

An example CUPL logic description file solution is shown in Fig. 14-4. A mod-7 ring counter will require 7 flip-flop stages in the shift register. To produce a single rotating 1 in the output of the shift register requires a 6-input NOR function for the feedback circuit. Inputs to the NOR will come from 6 out of the 7 flip-flop outputs. Only the last flip-flop stage in the serial shift register circuit will be omitted from the NOR's inputs. The feedback function needed for this ring counter is:

$$\overline{Q0 + Q1 + Q2 + Q3 + Q4 + Q5}$$

```
Name        RING-1;               Partno    L155-18C;
Date        07/21/93;             Revision  01;
Designer    G. Moss;              Company   Digi-Lab;
Assembly    Controller;           Location  U919;
Device      G16V8A;               Format    j;


/************************************************************/
/*    7-bit ring counter - output rotates a single 1    */
/************************************************************/


/**   Inputs   **/
PIN 1 = clock;                /* Shift register clock    */
PIN 11 = !oe;                 /* Register output enable  */


/**   Outputs   **/
PIN [12..18] = [Q6..0];         /* Register outputs    */
PIN 19 = feedbk;                /* Feedback signal     */


/**   Declarations & Intermediate Variable Definitions **/

field ring = [Q0..6];              /* Ring counter output */


/** Logic Equations **/

ring.d   =   [!feedbk, Q0, Q1, Q2, Q3, Q4, Q5];

feedbk   =   [Q0..5]:#;     /* OR function of Q0 thru Q5 */
```

Fig. 14-4 CUPL logic description file for example 14-3

117

This was implemented in the design file by first ORing the set of variables [Q0..5] to produce "feedbk" using the equality operation notation and then inverting "feedbk" in the logic equation list of bits for the "ring" output after the clock. The shifting operation is accomplished by the order of the listing of the bits in the logic equation for "ring." The data will move serially from the Q0 input end of the shift register toward the Q6 output end.

Laboratory Projects

14.1   Waveform pattern generator

Design a waveform pattern generator using a parallel-in, serial-out shift register. The shift register can be used to produce any desired 8-bit waveform pattern such as the one illustrated below. Use an IC counter chip to control the parallel loading/serial shifting sequence. Construct and test the logic circuit by comparing the CLK and OUT waveforms with a two-channel oscilloscope. Use the Shift/Load signal to trigger the scope. Note the change in the OUT waveform as the data input switches are changed. Hints: There will be 8 clock cycles needed for each 8-bit pattern sequence: 1 cycle to parallel load the 8-bit pattern and 7 more to serially shift the data toward Q_H for serial output. Use a mod-8 counter to control the loading and shifting sequence.

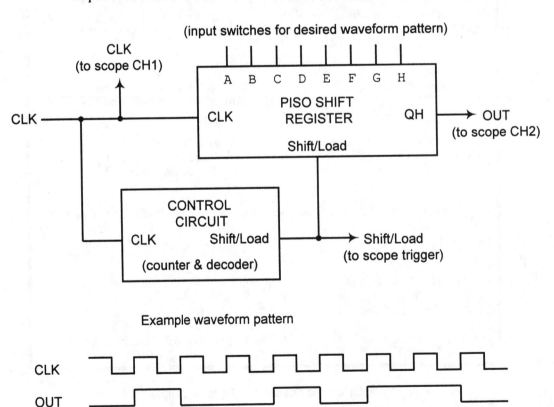

14.2   Serial data buffer

Design a serial data buffer using a serial-in, parallel-out shift register and a D-type latch. The circuit will accept 8 bits of serial data and then automatically transfer (in parallel) the data to a register. The register's output will be updated with new data after every 8 clock cycles. Use an IC counter chip to control the serial shifting/parallel transfer sequence. Construct and test the logic circuit using a <u>manual</u> clock signal (pushbutton) for CLK.

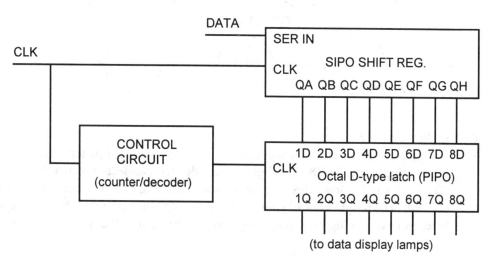

14.3   Mod-6 ring counter

Design a self-starting (and self-correcting), mod-6 ring counter using either a GAL16V8 or an appropriate MSI shift register chip(s) with any other necessary gates. The ring counter should rotate a single 0 for its sequence. Construct and test the counter circuit.

14.4   Mod-10 Johnson counter

Design a mod-10 Johnson counter using either a GAL16V8 or an appropriate MSI shift register chip(s) with any other necessary gates. The counter also needs to have an active-low reset control (RESET). Construct and test the counter circuit.

14.5 Shift register counter
Design a shift register counter with the sequence given in the following state transition diagram. Use a 74LS95B shift register chip.

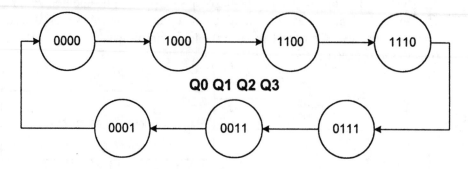

14.6 Light controller
Design a control circuit that will produce either of two light pattern sequences for 6 lights using a GAL16V8. The light pattern desired is selected with a signal called CTRL. An active-low RESET control input is used to initialize the pattern output to 000000. The 6 light control signals are labeled Q0 through Q5. The two sequences are listed below. The lights will light up one at a time until all 6 lights are turned on.

	CTRL = 0						CTRL = 1					
CLK	Q0	Q1	Q2	Q3	Q4	Q5	Q0	Q1	Q2	Q3	Q4	Q5
0	0	0	0	0	0	0	0	0	0	0	0	0
1	1	0	0	0	0	0	0	0	0	0	0	1
2	1	1	0	0	0	0	0	0	0	0	1	1
3	1	1	1	0	0	0	0	0	0	1	1	1
4	1	1	1	1	0	0	0	0	1	1	1	1
5	1	1	1	1	1	0	0	1	1	1	1	1
6	1	1	1	1	1	1	1	1	1	1	1	1
7	0	0	0	0	0	0	0	0	0	0	0	0
			repeats						repeats			

14.7   Parallel data pipeline

Design a 4-bit data pipeline circuit using a GAL16V8. A 4-bit data word (D3 D2 D1 D0) will be entered in parallel into the data pipeline when the LOAD control is high and the registers are clocked. The data pipeline is two words deep. The pipeline depth can be increased by cascading several of these chips together.

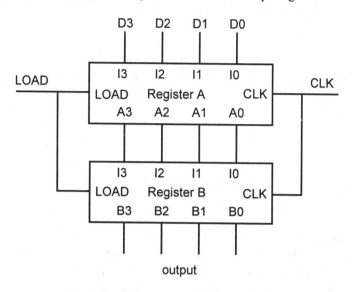

output

14.8   Data word logic unit

Design a logic circuit that can store two 4-bit data words in two 4-bit registers. The logic circuit can also generate and store a resultant data word that represents either the ANDing or XORing of the two words contained in the registers. Use a GAL16V8. Two function controls (F1, F0; see the table below) are used to select the desired function for the circuit. A control signal called REG-SEL determines which register (REG-A or REG-B) will be the destination of the operation. The register not specified as the destination will retain its current data. The data inputs are labeled I3, I2, I1, and I0.

F1	F0	Function
0	0	HOLD
0	1	LOAD
1	0	AND
1	1	XOR

REG-SEL	Register
0	REG-A
1	REG-B

14.9   Special-purpose data register
Design an 8-bit data register using a GAL16V8 that operates according to the following function table. The register function is selected with the controls S1 and S0. The 4-bit data input is labeled I3, I2, I1, and I0 and the register output is labeled Q7, Q6, Q5, Q4, Q3, Q2, Q1, and Q0.

S1	S0	Operation	$Q7_{n+1}$	$Q6_{n+1}$	$Q5_{n+1}$	$Q4_{n+1}$	$Q3_{n+1}$	$Q2_{n+1}$	$Q1_{n+1}$	$Q0_{n+1}$
			Register outputs after clock pulse							
0	0	Load L.S. nibble	$Q7_n$	$Q6_n$	$Q5_n$	$Q4_n$	$I3_n$	$I2_n$	$I1_n$	$I0_n$
0	1	Load M.S. nibble	$I3_n$	$I2_n$	$I1_n$	$I0_n$	$Q3_n$	$Q2_n$	$Q1_n$	$Q0_n$
1	0	Swap nibbles	$Q3_n$	$Q2_n$	$Q1_n$	$Q0_n$	$Q7_n$	$Q6_n$	$Q5_n$	$Q4_n$
1	1	Rotate data right	$Q0_n$	$Q7_n$	$Q6_n$	$Q5_n$	$Q4_n$	$Q3_n$	$Q2_n$	$Q1_n$

Note: L.S. = Least Significant
        M.S. = Most Significant

14.10   Data rotating register
Design a 6-bit data rotating register using a GAL16V8. The function of the register is to be controlled by inputs M2, M1, and M0 as shown in the following function table. The parallel data inputs are labeled D5 through D0 and the register outputs are Q5 through Q0. The data can be rotated either right (toward Q0) or left (toward Q5) in the register. A data rotation feeds the last data bit back around to the input of the shift register instead of losing it out the end.

M2	M1	M0	Function
0	0	0	HOLD DATA
0	0	1	LOAD DATA
0	1	0	ROTATE RIGHT
0	1	1	ROTATE LEFT
1	X	X	CLEAR REGISTER

Rotate direction	$Q5_{n+1}$	$Q4_{n+1}$	$Q3_{n+1}$	$Q2_{n+1}$	$Q1_{n+1}$	$Q0_{n+1}$
	Data rotation results					
RIGHT	$Q0_n$	$Q5_n$	$Q4_n$	$Q3_n$	$Q2_n$	$Q1_n$
LEFT	$Q4_n$	$Q3_n$	$Q2_n$	$Q1_n$	$Q0_n$	$Q5_n$

14.11   Digital lock

Design a digital lock circuit using a GAL16V8. The lock will have a 4-bit data input and an ENTER signal. Use a 4-bit shift register counter to keep track of the sequencing through the combination. The ENTER signal is the clock signal for the shift register. A 4-number lock combination sequence (established in the design) will be required to unlock the combination lock. Each input number (4-bit value) will be applied and then the ENTER signal will be asserted (active-high). The machine states must be sequenced in the proper order to unlock the lock. The "START" state, or STATE0, is followed by the intermediate states STATE1, then STATE2, then STATE3, and then finally STATE4, which "UNLOCKs" the lock. If an incorrect input combination is "ENTERed" during any counter state, the shift register counter will return to STATE0. The state transition diagram for the digital lock is shown below.

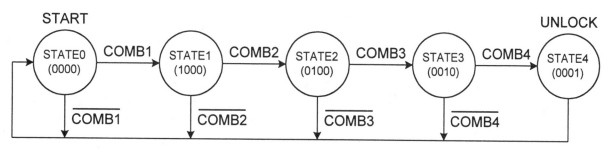

Returns to STATE0 if any incorrect combination is applied

14.12   Mod-31 LFSR counter

Design a mod-31 LFSR counter using a GAL16V8. The counter should have a synchronous, active-low RESET and an active-high count enable.

SYNCHRONOUS COUNTER DESIGN USING JK FLIP-FLOPS

Objectives

- To design a synchronous counter with a specified sequence using Karnaugh mapping and standard JK flip-flops.
- To verify synchronous counter designs using logic simulation software.

Suggested Parts				
74LS08	74LS32	74LS47	74LS112A	MAN72
Resistors: 330Ω				

Synchronous Counter Design

Synchronous sequential circuits may be designed by developing a transition table from the desired state sequence. A transition table is used to identify the synchronous inputs that must be applied to each flip-flop to produce the specified count sequence. The Boolean expression for each flip-flop input can be derived by Karnaugh mapping the transition table information. The set of logic equations describes the necessary input circuitry for each flip-flop. This procedure can be applied to any type of flip-flop and any desired count sequence (as long as the logic simplification is manageable with Karnaugh mapping).

Example 15-1

Design a synchronous, mod-7 counter using JK flip-flops that produces the sequence given in the state diagram in Fig. 15-1.

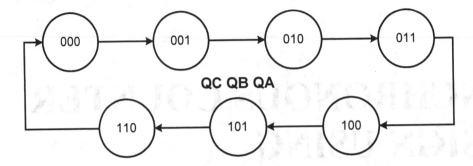

Fig. 15-1 Mod-7 counter state diagram for example 15-1

First, complete a present state - next state table, as illustrated in Table 15-1, to describe the desired sequence from the given state diagram.

Present State			Next State		
QC	QB	QA	QC	QB	QA
0	0	0	0	0	1
0	0	1	0	1	0
0	1	0	0	1	1
0	1	1	1	0	0
1	0	0	1	0	1
1	0	1	1	1	0
1	1	0	0	0	0
1	1	1	X	X	X

Table 15-1 Present state - next state table for example 15-1

Then determine the necessary flip-flop inputs to produce the indicated sequence and list the information in an excitation table (see Table 15-2).

Present States			State Transitions			Flip-flop Inputs					
QC	QB	QA	$QC_n \rightarrow QC_{n+1}$	$QB_n \rightarrow QB_{n+1}$	$QA_n \rightarrow QA_{n+1}$	JC	KC	JB	KB	JA	KA
0	0	0	0 → 0	0 → 0	0 → 1	0	X	0	X	1	X
0	0	1	0 → 0	0 → 1	1 → 0	0	X	1	X	X	1
0	1	0	0 → 0	1 → 1	0 → 1	0	X	X	0	1	X
0	1	1	0 → 1	1 → 0	1 → 0	1	X	X	1	X	1
1	0	0	1 → 1	0 → 0	0 → 1	X	0	0	X	1	X
1	0	1	1 → 1	0 → 1	1 → 0	X	0	1	X	X	1
1	1	0	1 → 0	1 → 0	0 → 0	X	1	X	1	0	X
1	1	1	1 → X	1 → X	1 → X	X	X	X	X	X	X

Table 15-2 Excitation table for example 15-1

Record the information for each flip-flop input in a separate K-map and determine the appropriate, simplified Boolean expressions. Note that "don't care" output conditions (Xs in the K-maps) may be defined as either 0s or 1s and may be used to simplify the expressions.

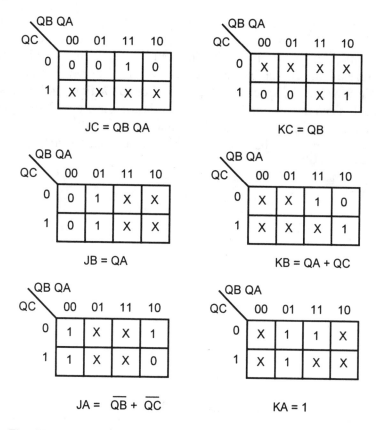

Fig. 15-2 Karnaugh mapping of J and K inputs for example 15-1

The schematic for the synchronous circuit design (see Fig. 15-3) can now be drawn from the J and K flip-flop input equations determined with K-mapping.

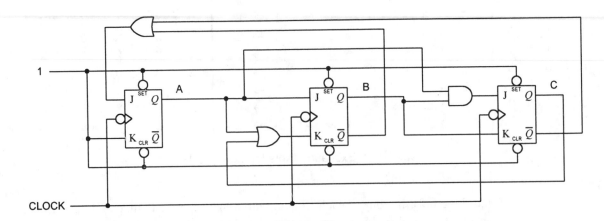

Fig. 15-3 Schematic for synchronous design in example 15-1

The circuit design should then be analyzed to verify proper circuit operation. It may also be necessary for the counter to be self-correcting for proper operation in the circuit application. If so, the circuit should be analyzed completely (all possible states) to determine if the circuit design is self-correcting. The self-correction feature can also be "designed in" by specifying the desired circuit action for the unused states in the present-state/next-state table initially. Analyzing this circuit design gives the state transition diagram shown in Fig. 15-4 and verifies correct circuit operation. Note that the circuit happens to be self-correcting since the unused state 111 returns to the proper sequence loop.

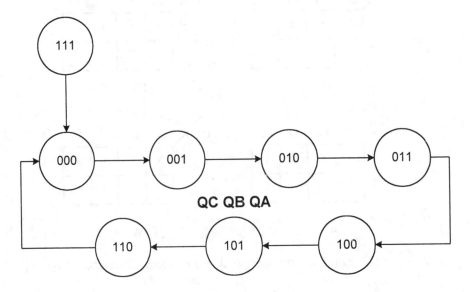

Fig. 15-4 Verifying circuit design in example 15-1

Stepper Motor Controller

Stepper motors produce rotary motion in discrete angular increments or steps. Stepper motors are used to feed the paper in a computer printer, control the 2-axis movement of XY plotters, control the movements of robotic arms, and control angular movement for many other types of automatic machines. The shaft of the stepper motor will rotate a specific number of degrees per "step." This is done by creating a magnetic field in the motor that can be rotated. This magnetic field will cause a permanent magnet attached to the rotor's shaft to rotate accordingly and "step the motor." To properly step the motor requires passing current through the appropriate stator windings (and, thereby, energizing them), which moves the shaft to the next angular position. Continued stepping is accomplished by energizing and de-energizing the stator windings of the motor in a specific sequence to create a rotating magnetic field in the motor. The stepper motor controller must produce the necessary sequence of states to control the stator windings. A 4-phase (4 stator windings) stepper motor can be made to rotate in either direction by controlling the windings according to the following state transition diagram shown in Fig. 15-5. The four stator windings are represented by the four variables W, X, Y, and Z. The direction of rotation (clockwise or counterclockwise) of the rotor is controlled by D. The stator windings will be energized when the appropriate controller output is high. Since the output current from a typical flip-flop is not great enough to power the motor, this will actually be accomplished by connecting the controller's outputs to a set of 4 stator drivers (such as transistors). The output of the controller then turns on the necessary drivers to energize the stator windings.

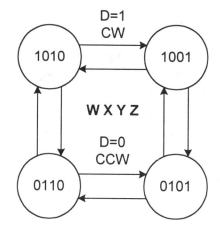

Fig. 15-5 Stepper motor control sequence

Laboratory Projects

Design synchronous counters using JK flip-flops for each of the given count sequences. Do not use the asynchronous flip-flop inputs in the circuit designs. Verify that each circuit design should produce the required sequence and determine if each circuit is self-correcting by analyzing the circuit designs. Construct and test each circuit design. Use an oscilloscope to display the timing diagram for each counter.

15.1   Mod-6 synchronous counter design

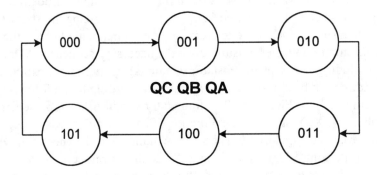

15.2   Mod-5 synchronous down counter design

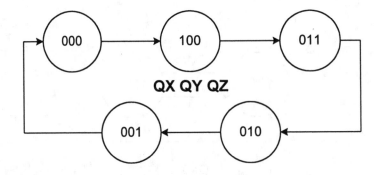

15.3   Mod-5 synchronous counter design

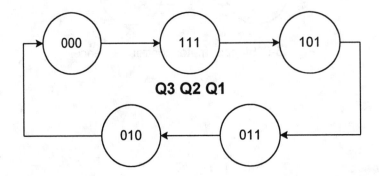

15.4 Synchronous, recycling 5421 BCD counter design

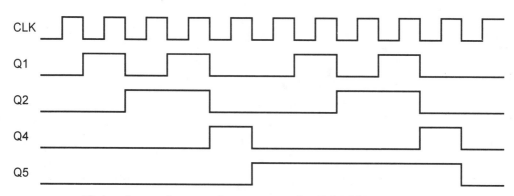

Timing diagram for a synchronous, recycling 5421 BCD counter

15.5 Synchronous 8421 BCD counter design
Connect the BCD counter to a 7-segment display using a 74LS47 decoder/driver.

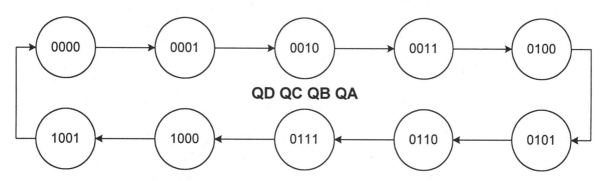

15.6 Stepper motor controller
Design a 4-phase, bidirectional stepper motor control circuit. The motor direction is to be controlled by D (D = 1 for CW and D = 0 for CCW). Remember, if you actually connect the controller to a stepper motor, use drivers for each of the stator windings. Hints: This circuit can be designed using a total of two chips: one with 4 XOR gates and the other with 2 JK flip-flops. The four signals W, X, Y, and Z can be produced from the two JKs since the output pattern always has:

$$W = \overline{X} \quad \text{and} \quad Y = \overline{Z}$$

DEVICE CHARACTERISTICS

Objectives

- To measure the power requirements for logic chips.
- To measure propagation delays of logic gates.
- To measure the output current capabilities of a logic gate.
- To measure the logic level input voltage thresholds for logic gates.
- To measure the transfer characteristic of logic gates.

Suggested Parts			
74HC00	74LS00	74LS14	potentiometers: 10 KΩ, 50 KΩ

Device Characteristics

For proper operation, digital chips must meet standardized parametric specifications. Parametric testing of various digital device voltage, current, and time delay characteristics may be performed to ensure that a device meets these specifications.

Chip Power Supply Current Requirements

The power requirements for a logic chip are determined by measuring the amount of current drawn by the chip from its power supply. Since an individual gate's current needs may vary depending on its output logic level, the average current is often specified. The supply current for SSI chips is measured with all gate outputs at a high logic level (I_{CCH} or I_{DDH}) and again with the outputs at a low logic level (I_{CCL} or I_{DDL}). An average current is then computed. The simplest technique would be to use a digital multimeter (DMM) to measure the supply currents for the two output conditions for the device under test (DUT). Tie the gate inputs to the appropriate logic levels to produce the two specified static gate output conditions. The supply current level for CMOS chips (I_{DD}) is dependent on the frequency of the input signals. This may be tested by applying a compatible square wave signal to the gate inputs and measuring I_{DD}. The current reading will automatically be averaged with the 50 percent duty cycle square wave input.

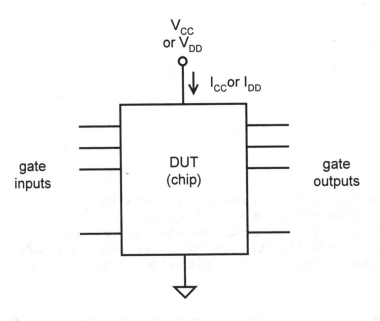

Fig. 16-1 Measuring supply current

Propagation Delays

Propagation delay times are measured from the 50 percent transition point on the input waveform to the 50 percent transition point on the resulting output waveform. The two propagation delay times specified for a gate output are called t_{PLH} and t_{PHL} and are referenced, respectively, to the low-to-high and high-to-low transitions of the output waveform. Propagation delay times are difficult to measure since they are typically only a few nanoseconds and it is best to use a high bandwidth scope (>100MHz) for more precise measurements. However, it is possible to cascade several gates together to measure an overall propagation delay and then calculate an average delay per gate. The input signal should be a voltage-compatible square wave of around 1MHz.

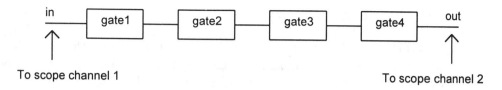

Fig. 16-2 Measuring average propagation delays

Output Current Sinking and Sourcing

The output current drive capability of a logic gate limits the number of gate inputs that can be connected to a single output. This loading capacity is referred to as the fan-out of a logic gate. The current-sinking (I_{OL}) and current-sourcing (I_{OH}) ability of a logic gate can be measured as shown in Fig. 16-3. Apply the appropriate input signals to produce the desired high or low output level from the gate (DUT) while the variable resistor (use approximately 1K ohms for TTL gates and 50K ohms for CMOS gates) is at its maximum resistance. Then adjust the variable resistor to increase the output current until the measured output voltage reaches the specified V_{OLmax} or V_{OHmin}. Measure the maximum output current levels.

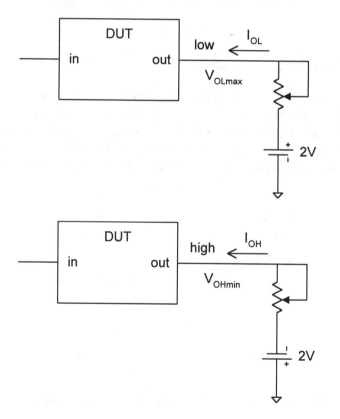

Fig. 16-3 Measuring output current drive

Input Voltage Levels

Logic level voltages for gates are defined by the triggering threshold values that cause the gate outputs to change states. Gate triggering thresholds can be measured by varying the input voltage from 0 volts to the chip's supply voltage value and monitoring the gate output voltage for the specified V_{OLmax} and V_{OHmin}. Use a DMM to measure V_{in} and V_{out}. Gate outputs will often oscillate when the input voltage is between V_{ILmax} and V_{IHmin}.

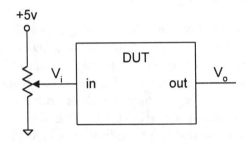

Fig. 16-4 Measuring input threshold voltages

Gate Transfer Characteristics

The transfer characteristic for a device is the relationship between its input and output signals. The transfer characteristic can be obtained graphically using an oscilloscope in the X-Y mode. A triangle (or sine) wave signal is used to sweep the device's input between the two extremes (ground and V_{CC}). The DUT input voltage signal is monitored on the X (horizontal) input and the DUT output voltage signal is monitored on the Y (vertical) input of the oscilloscope.

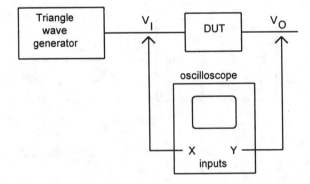

Fig. 16-5 Measuring device's transfer characteristic

Laboratory Projects

16.1   Logic chip power requirements
Use a DMM to measure the average supply current for a typical TTL logic chip (such as a 74LS00) and a functionally equivalent CMOS chip (such as a 74HC00). Measure the supply current for the following gate input conditions: static inputs, 1KHz, and 1MHz square wave (TTL compatible) input signals.

16.2   Gate propagation delay times
Use an oscilloscope to measure the propagation delays for a typical TTL logic chip (such as a 74LS00) and a functionally equivalent CMOS chip (such as a 74HC00).

16.3   Gate output current capabilities
Use a DMM to measure the output current sourcing and sinking ability (I_{OH} and I_{OL}) for a typical TTL logic chip (such as a 74LS00) and a functionally equivalent CMOS chip (such as a 74HC00).

16.4   Gate input voltage levels
Use a DMM to measure the input voltage levels V_{ILmax} and V_{IHmin} for a typical TTL logic chip (such as a 74LS00) and a functionally equivalent CMOS chip (such as a 74HC00).

16.5   Schmitt trigger transfer characteristic
Use an oscilloscope (in X-Y mode) to obtain the transfer characteristic of one of the Schmitt trigger inverters in a 74LS14. **Be sure to adjust the triangle or sine wave generator signal using the oscilloscope to $0 \leq V_{in} \leq +5v$ <u>before</u> applying it to the gate's input.**

DECODERS AND DISPLAYS

Objectives

- To apply decoding circuits in digital system applications.
- To implement digital displays using 7-segment devices.

Suggested Parts					
74LS00	74LS47	74LS138	74LS393	GAL16V8	MAN72
Resistors: 330 Ω					

Decoder Circuits

A decoder is used to detect a particular combination of bits applied to the input of the circuit and to display that information in a specified fashion. Logic gates can be used to design any type of decoder circuit. Some decoders are available as IC chips. For example, the 74LS138 shown in Fig. 17-1 is a 3-line to 8-line decoder. It can be used to identify which of 8 possible input combinations is applied to its 3 data input lines. The docoder's 8 outputs are active-low. This chip also has 3 enable inputs (2 active-low and 1 active-high), which adds greatly to its flexibility and usefulness.

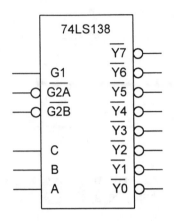

Fig. 17-1 74LS138, 3-line to 8-line decoder chip

Decoder/Drivers and 7-Segment Displays

Decoder/driver circuits are available for various kinds of display devices such as 7-segment displays. A 7-segment display device is commonly used to display the decimal characters 0-9. The display segments are often constructed using light emitting diodes (LEDs) in which the appropriate LED segments are forward-biased (causing them to emit light) for the desired symbol shape. Decoder/driver circuits control the LED biasing for the display of the appropriate characters for the data being input. Series current limiting resistors are employed to protect the individual LED segments from damage caused by too much forward-biased diode current. The pin-out configuration for a typical common-anode, 7-segment display is illustrated in Fig. 17-2.

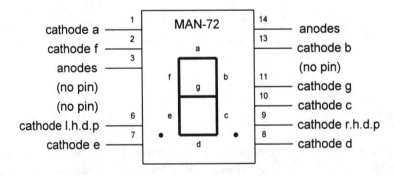

Fig. 17-2 MAN72 (or equivalent), 7-segment, common-anode display pin assignments (top view)

Standard BCD-to-7-segment decoder/driver chips are available to provide the necessary biasing signals for a 7-segment LED display device to produce the decimal characters of 0 through 9. Since common-anode and common-cathode 7-segment devices are available, an appropriate decoder/driver chip must be selected to match the display type. The 74LS47 shown in Fig. 17-3 is designed to be used with a common-anode type device. Note that a series resistor is needed for each segment of the display to limit the amount of LED current to a safe level.

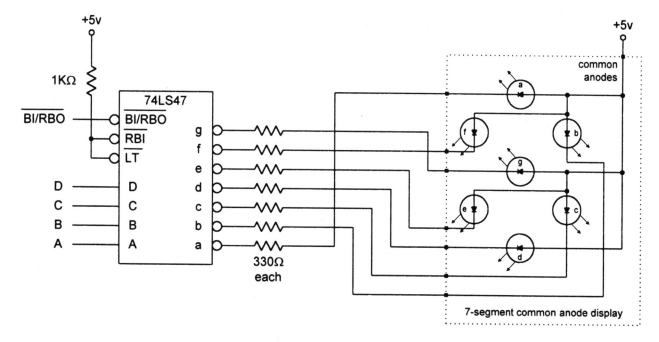

Fig. 17-3 Standard decoder/driver and 7-segment LED display circuit

Example 17-1

Design a custom decoder/driver circuit using a GAL16V8 that can be used to produce the display patterns on a 7-segment, common-anode, LED display shown in Table 17-1. The inputs for the decoder and display circuit are E, C, B, and A.

E	C	B	A	Display
1	0	0	0	A
1	0	0	1	b
1	0	1	0	C
1	0	1	1	d
1	1	0	0	E
1	1	0	1	F
1	1	1	0	H
1	1	1	1	L
0	X	X	X	blank

Table 17-1 7-segment display patterns for example 17-1

A possible solution source file is shown in Fig. 17-4. A truth table design entry technique was chosen for its ease in defining this project. A "1" in the segment output list indicates that that particular output should be "active." The active level for each of the segment outputs is given in the pin assignment statement. For common-anode, 7-segment displays, the segment inputs are all active-low.

```
Name            CHARACTR;
Partno          L105-22A;
Date            07/13/93;
Revision        01;
Designer        Greg Moss;
Company         Digi-Lab, Inc.;
Assembly        display board;
Location        U612;
Device          G16V8A;
Format          j;

    /************************************************/
    /*    Decoder/driver for C.A. 7-segment display  */
    /*        produces specified characters          */
    /************************************************/


/** Inputs  **/

pin [2..5] = [E,C,B,A];              /* Enable and
                                          Data inputs */


/** Outputs **/

pin [19..13] = ![a,b,c,d,e,f,g];   /* Segment outputs */


/**          Declarations and Intermediate
                  Variable Definitions              **/

field input = [E,C,B,A];     /* Hexadecimal input field */
field segment = [a,b,c,d,e,f,g]; /* Display segments   */


/** Logic Equations **/

table           input  =>     segment          {
     /*                       abcdefg                   */
                  8    =>    'b'1110111;
                  9    =>    'b'0011111;
                  A    =>    'b'1001110;
                  B    =>    'b'0111101;
                  C    =>    'b'1001111;
                  D    =>    'b'1000111;
                  E    =>    'b'0110111;
                  F    =>    'b'0001110;
                                               }
```

Fig. 17-4 CUPL source file for example 17-1

Example 17-2

Design a memory chip decoder using a GAL16V8. The decoder should produce individual chip select signals for three memory chips. The decoder should also be enabled by two active-low, memory control, strobe signals. One of the memory chips will be enabled by only one of the strobe signals, while the other two memory chips can be enabled with either strobe signal. The two strobe signals cannot be active simultaneously. The decoder's function table and specific address range for each memory chip is given in Table 17-2. The 16-bit addresses are given in hexadecimal in the table. Only the upper 5 bits of the address (A15 through A11) will be used to decode the information for each memory chip in this application.

An example source file solution is given in Fig. 17-5. Each chip select signal will be active when the appropriate address range is present and the proper enable signals are active. Only the address inputs that distinguish the address ranges from each other need to be decoded. The active levels for the inputs and outputs are defined in the pin assignment statements. The variable "memreq" was defined to indicate when either of the two strobe signals (memw or memr) is active.

Address Range (hexadecimal)	Memory Strobes		Chip Selects		
	memw	memr	cs2	cs1	cs0
0000-1FFF	1	0	0	1	1
0000-1FFF	1	1	1	1	1
2000-27FF	1	0	1	1	0
2000-27FF	0	1	1	1	0
2000-27FF	1	1	1	1	1
2800-2FFF	1	0	1	0	1
2800-2FFF	0	1	1	0	1
2800-2FFF	1	1	1	1	1

Table 17-2 Decoder function table for example 17-2

Example 17-3

Design a 1-out-of-4 decoder using a GAL16V8. The decoder will have active-low outputs Y3 Y2 Y1 Y0 and an active-low enable named EN. The inputs are IN1 IN0.

An example CUPL solution using the $repeat preprocessor command is shown in Fig. 17-6. The CUPL statement in the body is repeated 4 times using the index variable i. The decoder output that is asserted will be dependent upon the input value that is controlled by the index value. The repeat body is ended with the $repend preprocessor command.

```
Name       CHIPSLCT;              Partno    L105-22B;
Date       07/13/93;             Revision  01;
Designer   Greg Moss;            Company   Digi-Lab, Inc.;
Assembly   Memory board;         Location  U702;
Device     G16V8A;               Format    j;

/*******************************************************/
/* decoder generates chip select signals for two memory */
/*    chips (one 8Kx8 ROM and two 2Kx8 static RAMs).   */
/*******************************************************/

/**  Inputs  **/
PIN [2..6]   = [A15..11];      /* Address Bus Lines    */
PIN [7,8]    = ![memw,memr];   /* Memory Data Strobes  */

/**  Outputs  **/
PIN [19..17] = ![cs2..0];      /* Chip Selects  */

/** Declarations and Intermediate Variable Definitions **/
field memadr = [A15..11];      /* Address named "memadr"  */
memreq = memw # memr;          /* define variable "memreq" */

/**  Logic Equations  **/
cs2 = memr    &  memadr:[0000..1FFF];
cs0 = memreq  &  memadr:[2000..27FF];
cs1 = memreq  &  memadr:[2800..2FFF];
```

Fig. 17-5 CUPL source file for example 17-2

```
Name       decode4;               Partno    L109-23;
Date       12/08/96;             Rev       01;
Designer   Greg Moss;            Company   Digi-Lab, Inc.;
Assembly   Decoder Board;        Location  U138;
Device     G16V8A;        Format    j;
/***************************************************/
/* 1-out-of-4 decoder with active-low enable     */
/***************************************************/

/** Inputs **/
Pin 1 = !EN;                   /* active-low enable  */
Pin [2,3] = [IN1..0];          /* data inputs        */

/** Outputs **/
Pin [12..15] = ![Y0..3];   /*  outputs            */

/** Declarations & Intermediate Variable Definitions **/
field DATA = [IN1..0];       /* data inputs         */

/** Equations **/
$repeat i=[0..3]
  Y{i} = DATA:{i} & EN;
$repend
```

Fig. 17-6 CUPL source file for example 17-3

Example 17-4

Design a binary-to-BCD code converter using a GAL16V8. The code converter will have a 5-bit input (00000_2 through 11111_2) and a 2-digit output (00 through 31_{10} in BCD).

An example source file solution is given in Fig. 17-7. Two truth tables are used to define the desired outputs for each of the two digits. The most significant digit output will be determined by the binary input being less than 10, in the teens, in the twenties, or in the thirties. A range of values is given in list notation for each of the four output conditions. The least significant digit is determined by the least significant digit of the input value. The set of appropriate input values for each output condition is given in list notation. The quantities are all expressed in decimal for ease of interpretation.

```
Name          bin-BCD;          Partno       L159-17-4;
Date          12/05/96;         Revision     01;
Designer      Greg Moss;        Company      Digi-Lab;
Assembly      Decoder board;    Location     U117;
Device        G16V8A;           Format       j;
/*********************************************************/
/*   design to convert 5-bit binary to 2-digit BCD   */
/*********************************************************/

/**   Inputs  **/
Pin [1..5]   = [I4..0];    /*  Binary input  */
/**   Outputs  **/
Pin [12..15] = [L0..3];    /*  Least significant BCD output */
Pin [16..19] = [M0..3];    /*  Most significant BCD output  */

/** Declarations & Intermediate Variable Definitions **/
field BIN  = [I4..0];    /*  Binary input  */
field LSD  = [L3..0];    /*  LSD output    */
field MSD  = [M3..0];    /*  MSD output    */

/**   Truth Tables  **/
table  BIN      =>    MSD    {
  'd'[0..9]    =>    'd'0;
  'd'[10..19]  =>    'd'1;
  'd'[20..29]  =>    'd'2;
  'd'[30..31]  =>    'd'3;    }

table    BIN        =>    LSD    {
 'd'[0,10,20,30]  =>    'd'0;
 'd'[1,11,21,31]  =>    'd'1;
 'd'[2,12,22]     =>    'd'2;
 'd'[3,13,23]     =>    'd'3;
 'd'[4,14,24]     =>    'd'4;
 'd'[5,15,25]     =>    'd'5;
 'd'[6,16,26]     =>    'd'6;
 'd'[7,17,27]     =>    'd'7;
 'd'[8,18,28]     =>    'd'8;
 'd'[9,19,29]     =>    'd'9;    }
```

Fig. 17-7 CUPL source file for example 17-4

Laboratory Projects

Design decoder and display circuits for the following applications. Breadboard and verify your designs.

17.1   4-bit counter and display

Connect a 7-segment display to one of the 4-bit counters in a 74LS393 using a 74LS47. Design a control circuit for the decoder/driver so that only valid BCD characters (0 through 9) will be displayed, even though the 4-bit counter will have a count sequence of 0 through 15 in binary. The display should blank out the invalid BCD characters produced by the 74LS47 when its data input is greater than 9. Also display the binary output of the counter on LEDs so that you can monitor the count progress even while the display is blanked. Use a 74LS393 counter chip; a 74LS47 decoder/driver chip; a common-anode, 7-segment display; and <u>one</u> additional SSI chip. **Remember to use a series current limiting resistor (approximately 330 ohms) for <u>each</u> LED segment.**

17.2   Octal counter decoder and display

Use a 74LS138 to decode 4 specified states (see table below) that are output by an octal counter. We will assume that the decoded states are to be used to provide control signals for 4 different circuits (W, X, Y, or Z) as indicated in the truth table. The table indicates that the decoder is also controlled by two enable signals (EN2 and EN1). The outputs QC QB QA from a 74LS393 counter will be used to generate the octal count sequence. Continuously monitor the counter state (0 through 7) by <u>also</u> connecting the counter's outputs to a 7-segment display <u>with</u> a standard 7-segment decoder/driver.

| Enables | | Counter | | | Decoder | | | | 7-segment |
EN2	EN1	QC	CB	QA	W	X	Y	Z	Display
1	0	0	0	0	0	1	1	1	0
1	0	0	0	1	1	1	1	1	1
1	0	0	1	0	1	1	1	1	2
1	0	0	1	1	1	0	1	1	3
1	0	1	0	0	1	1	1	1	4
1	0	1	0	1	1	1	0	1	5
1	0	1	1	0	1	1	1	1	6
1	0	1	1	1	1	1	1	0	7
0	X	X	X	X	1	1	1	1	counter state
X	1	X	X	X	1	1	1	1	counter state

17.3   Mod-16 counter and decoder

Design a decoder for a mod-16 binary counter (74LS393) using 74LS138 decoder chips. The decoder outputs need to be active-low and it also needs to have an active-low enable named E.

17.4 Hex decoder/driver

Design a hexadecimal decoder/driver circuit using a GAL16V8 that will drive a common-anode, 7-segment LED display. The resultant display output should be the equivalent hexadecimal value of 0 through F. The input to the decoder/driver is a 4-bit binary value produced by a mod-16 binary counter and the output is a set of 7 lines that are connected to the 7-segment display. To prevent ambiguous display characters, use lower case "**b**" (for 1011) and "**d**" (for 1101) and light the a-segment for a **6** (0110). **Remember to use a series current limiting resistor (approximately 330 ohms) for <u>each</u> LED segment.**

17.5 Binary-to-BCD converter and display

Design a binary-to-BCD converter using a GAL16V8. The converter should output the equivalent BCD value for a 6-bit binary input. Display the converter output on a 2-digit, 7-segment display. The converter can handle numbers from 0_{10} through 63_{10}. Blank a leading 0 (in the ten's digit) in the display.

17.6 Memory decoder

Design a memory chip decoder using a GAL16V8. The decoder should produce individual chip select signals for six memory chips. The decoder should also be enabled by either of two active-low memory control strobe signals (memr and memw). The two strobe signals cannot be active simultaneously. The specific address range to be decoded for each memory chip is given below. The 16-bit addresses are given in hexadecimal in the table. Hint: You will need to decode only the upper 6 bits of the address (A15 through A10).

Address Range (hexadecimal)	Chip Selects					
	CS5	CS4	CS3	CS2	CS1	CS0
0000–03FF	1	1	1	1	1	0
0400–07FF	1	1	1	1	0	1
0800–0BFF	1	1	1	0	1	1
0C00–0FFF	1	1	0	1	1	1
A800–ABFF	1	0	1	1	1	1
AC00–AFFF	0	1	1	1	1	1

17.7 Mod-100 BCD counter and 7-segment display

Design a 2-digit, 7-segment display for a mod-100 BCD counter. Blank the tens digit display if it is zero.

17.8 1-out-of-8 decoder

Design a 1-out-of-8 decoder using a GAL16V8. The decoder will have active-low outputs Y7 through Y0 and two enables, one active-low enable named EN0 and a second active high enable named EN1. The inputs are IN2 IN1 IN0.

UNIT 18

ENCODERS

Objective

- To apply encoder circuits in digital system applications.

Suggested Parts				
74LS47	74LS148	74LS393	GAL16V8	MAN72
Keypad or SPST switches		Resistors: 330 Ω, 1 KΩ		

Encoder Circuits

Input data to a logic system often comes from switches. This switch input information must be encoded into a representative binary form that can be processed by the system. Encoders have several data lines for the inputs that are being applied. The encoded output will be a binary number that will be used by the digital system to represent the specific input condition. Simple types of encoders can handle only one active input at a time. A priority encoder has an established order of precedence for the input lines so that the input with the highest priority will be encoded when multiple inputs are active simultaneously. MSI encoder chips such as the 74LS148 (see Fig. 18-1) are available. For added flexibility, this priority encoder chip has an enable input and enable and strobe outputs.

149

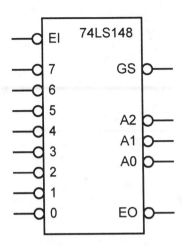

Fig. 18-1 74LS148, 8-line to 3-line priority encoder chip

Scanning Encoder

A scanning encoder can be utilized when there are many switch inputs to be encoded in a digital system. Standard functional blocks such as counters, decoders, and encoders can be combined to implement a scanning encoder. The schematic in Fig. 18-2 illustrates a scanning encoder system for a matrix of 64 SPST pushbutton switches. The scanning encoder will output a 6-bit binary number for $Z5$ $Z4$ $Z3$ $Z2$ $Z1$ $Z0$ that is equivalent to a pressed switch's decimal label.

The scan counter (74LS163) is used as a 3-bit binary counter that controls the repetitive scanning of the 8 rows of switches. The row scanning decoder (74LS138) decodes the scan counter's output to determine which switch matrix row is currently being scanned. Only one row at a time is scanned with the active-low output produced by the scanning decoder. A closed switch will short one of the columns to one of the rows. Each switch represents a unique combination of a row and column. A closed switch on a particular row is indicated by a low input to the 8-input column encoder (74LS148) <u>when</u> the appropriate row with the pressed switch is scanned with a low from the decoder. When the closed switch is detected by a low input to the encoder, the GS output (Scan Disable) of the encoder will go low. A low on Scan Disable will stop the counter (and the scanning of the switch rows) so that the binary code for the switch may be read out as the 6 bits $Z5$ $Z4$ $Z3$ $Z2$ $Z1$ $Z0$. Note that inverters are used to change the inverted binary encoder output into the standard binary value for the corresponding column.

As long as there is no pressed switch in the row that is currently being scanned, all columns will be pulled high with the resistors. With all highs into the encoder the GS output (Scan Disable) will be high and the counter will be enabled. This allows the counter to step to the next state when it is clocked and the decoder output will then change to scan the next row. Any pressed switch in a row that is <u>not</u> currently being scanned (with a low from the decoder) will have no effect on the encoder or counter since both sides of the switch will be high. The counting and row scanning will proceed until a closed switch is detected in the currently scanned row. A pressed switch will only stop the counter when that specific row is scanned. Releasing the switch will then allow the count to continue. If switch "21" (located in row 5) is pressed, column 2 will go low when the counter reaches state 101 ($Z_2 Z_1 Z_0$). The low on input 2 of the encoder will produce a 010 for $Z_5 Z_4 Z_3$. This 6-bit number 010101 is the binary equivalent of the decimal number 21.

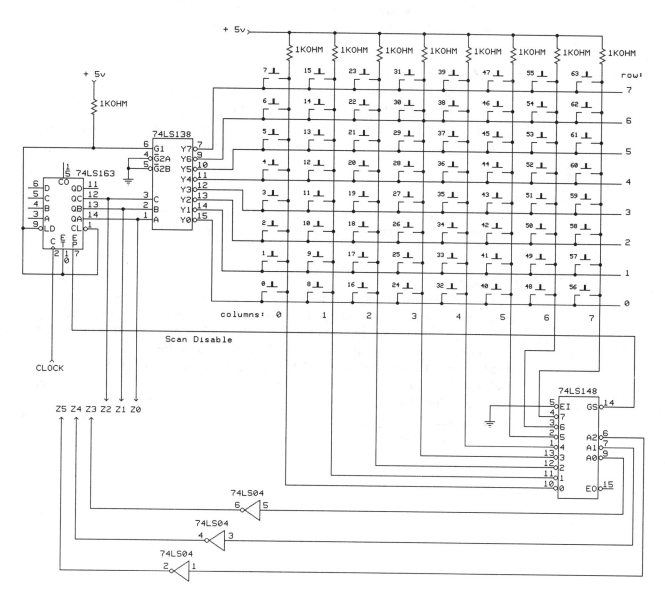

Fig. 18-2 Scanning encoder system

Laboratory Projects

Design encoder circuits for the following applications. Breadboard and verify your designs.

18.1   Encoder and display
Design an octal-to-binary switch encoder and 7-segment display circuit using a 74LS148 priority encoder chip, a 74LS47 decoder/driver chip, and any other necessary devices. The display should be blanked when there is no switch input being applied.

18.2   BCD encoder
Design a decimal-to-BCD encoder using a GAL16V8. The ten inputs are all active-low. The encoder should produce a standard 8421 BCD output corresponding to a single active input. A strobe signal (S) with a high output whenever one of the ten inputs is applied should also be produced by the encoder.

18.3   16-line priority encoder
Implement a 16-line priority encoder using two 74LS148 encoder chips and any other necessary devices. Hint: See the application notes for this chip in the appendix.

18.4   Scanning encoder
Implement the 16-input scanning encoder design illustrated in Fig. 18-3 with a GAL16V8 and a 74LS160 counter chip. The PLD can be used for the row scanning decoder, column encoder, and scan counter controller blocks indicated in the diagram. Connect a hexadecimal keypad (4 x 4 pushbutton switches) for inputs to the scanning encoder or use 4 toggle switches (SPST) to test the encoder for selected switch inputs (such as 0, 7, 10, and 15). Hints: The easiest design solution for the PLD is to write logic expressions for the seven outputs given in the three function tables. Remember to define the active levels in the pin assignment statements and then to write the equations in terms of when the outputs should be active.

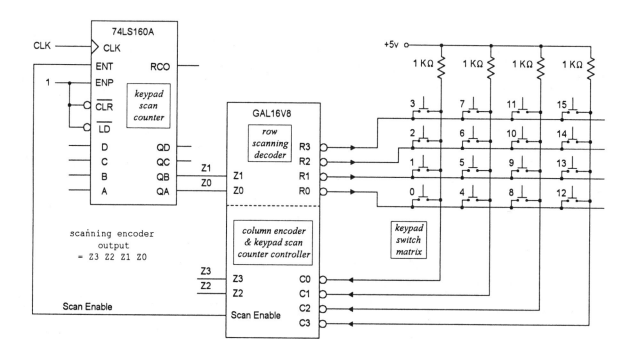

Row scanning decoder

Z1	Z0	R0	R1	R2	R3
0	0	0	1	1	1
0	1	1	0	1	1
1	0	1	1	0	1
1	1	1	1	1	0

Column encoder

C0	C1	C2	C3	Z3	Z2
0	1	1	1	0	0
1	0	1	1	0	1
1	1	0	1	1	0
1	1	1	0	1	1

Scan counter controller

C0	C1	C2	C3	Scan Enable
0	X	X	X	0
X	0	X	X	0
X	X	0	X	0
X	X	X	0	0
1	1	1	1	1

Fig. 18-3 Schematic & function tables for a scanning encoder system using a PLD

MULTIPLEXERS

Objective

- To apply multiplexer circuits in digital system applications.

Suggested Parts					
74LS04	74LS08	74LS47	74150	74LS151	74LS393
GAL16V8	MAN72	Resistors: 330 Ω			

Multiplexer Circuits

A multiplexer (also called a data selector) is a logic block that steers selected input data to its output. This logic function has multiple data inputs from which to choose, but only one of the inputs will appear on the output. Control signals called select lines are used to determine which data input will be routed to the output. Because of its ability to output different data inputs dependent on a set of control lines, a multiplexer can also easily be used to synthesize a combinational logic function. MSI multiplexer chips such as the 74LS151 (see Fig. 19-1) are available. This chip is a one-of-eight or 8-channel multiplexer. Three select lines are necessary to choose one of the eight possible inputs for the data output. For added flexibility, this chip has both a normal data output and an inverted data output, as well as a strobe or enable control input.

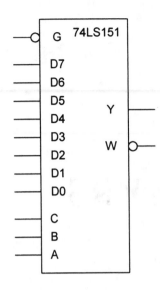

Fig. 19-1 74LS151, 8-channel multiplexer chip

Example 19-1

Design a dual, 4-channel multiplexer using a GAL16V8. The function table for this circuit is listed in Table 19-1. The 2-bit data input words are labeled A, B, C, and D and the 2-bit output word is Y. An active-low enable control is labeled EN and the two selection controls are labeled S1 and S0.

EN	S1	S0	Y1	Y0
0	0	0	A1	A0
0	0	1	B1	B0
0	1	0	C1	C0
0	1	1	D1	D0
1	X	X	0	0

Table 19-1 Function table for example 19-1

A CUPL source file solution is shown in Fig. 19-2. Variable names have been assigned (using the "field" statement) to each of the data input and output words and also to the select control lines. Each of the select code combinations has also been assigned logical variable names using the equality operator. One equation statement can be used to describe the function of both output bits (Y1 and Y0) simultaneously. CUPL will automatically create the appropriate equations with the correct "subscripts" for the two multiplexer outputs.

```
Name        DUAL_4CH;            Partno    L105-24;
Date        07/15/93;            Revision  01;
Designer    Greg Moss;           Company   Digi-Lab, Inc.;
Location    U901;                Assembly  mux board;
Device      G16V8A;              Format    j;

/************************************************/
/*  Dual 4-to-1 MUX                             */
/*      selects one of four 2-bit words         */
/************************************************/


/**    Inputs    **/
Pin  [9,11]   =   [S0,S1];     /* Select controls  */
Pin  [1,2]    =   [A1,A0];     /* Word A inputs     */
Pin  [3,4]    =   [B1,B0];     /* Word B inputs     */
Pin  [5,6]    =   [C1,C0];     /* Word C inputs     */
Pin  [7,8]    =   [D1,D0];     /* Word D inputs     */
Pin   19      =     !EN;       /* Enable control    */


/**    Outputs   **/
Pin [15,16]  =   [Y0,Y1];     /* MUX outputs        */


/** Declarations and Intermediate Variable Definitions **/

field select = [S1,S0];     /* select control   */
field A_IN = [A1,A0];       /* data inputs       */
field B_IN = [B1,B0];
field C_IN = [C1,C0];
field D_IN = [D1,D0];

field OUTPUT  = [Y1,Y0];    /* data outputs      */

SEL_A = select:0;           /* data selects      */
SEL_B = select:1;
SEL_C = select:2;
SEL_D = select:3;


/**   Logic Equations   **/

OUTPUT  =   (A_IN  &   SEL_A
        #    B_IN  &   SEL_B
        #    C_IN  &   SEL_C
        #    D_IN  &   SEL_D)   &   EN;
```

Fig. 19-2 CUPL design source file for example 19-1

Laboratory Projects

Design multiplexer circuits for the following applications. Breadboard and verify your designs.

19.1   Multiplexed BCD display

Design a single-digit, 7-segment display multiplexer using a GAL16V8 and a 74LS47 decoder/driver chip. Use logic switches for one BCD input and the 74LS393 counter for the other BCD input to the MUX. Only one of the two 4-bit numbers input to the multiplexer will be displayed on the 7-segment display as indicated by the following function table. Use the PLD to provide also an error function E that indicates if an invalid BCD input is applied to the logic switches. The gating control G will produce an invalid BCD output (1111) from the MUX when it is disabled. Hints: Use the PLD to implement a quad 2-line-to-1-line MUX. The counter can be cleared (or reset) on the state 1010 with a decoding gate to make the counter appear to produce a BCD output sequence.

		MUX Outputs			
G	S	Y3	Y2	Y1	Y0
1	0	SW3	SW2	SW1	SW0
1	1	QD	QC	QB	QA
0	X	1	1	1	1

19.2   Dual waveform pattern generator

Design a logic circuit using a 74LS151, one-of-eight multiplexer that will output either of two possible recycling waveform patterns selected by F. Use a 3-bit binary counter (74LS393) to sequence the multiplexer. View the CLOCK and OUT waveforms with a dual trace oscilloscope. Hint: Use the 3-bit counter to select the appropriate data value to be output in the sequence.

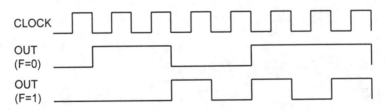

19.3   Prime number detector using a multiplexer
Implement a logic function generator using the 74150 multiplexer chip that will detect 5-bit prime numbers. The output should be high only if a prime number (1, 2, 3, 5, 7, 11, 13, 17, 19, 23, 29, or 31) is input to the detector. Hint: With 5 input bits to the detector and only 4 select lines on the MUX, some of the data inputs to the MUX can be variable instead of fixed and dependent on the fifth bit.

19.4   Quad 3-channel MUX
Using a GAL16V8, design a multiplexer that can select one of three 4-bit binary words to be output. The inputs are A3 A2 A1 A0, B3 B2 B1 B0, and C3 C2 C1 C0. The multiplexer's outputs are Y3 Y2 Y1 Y0 and the select controls are S1 S0. The function table for this MUX is given below. The output from the MUX will be 0000 when S1 = S0 = 0 (i.e., the MUX is disabled).

S1	S0	Y3	Y2	Y1	Y0
0	0	0	0	0	0
0	1	A3	A2	A1	A0
1	0	B3	B2	B1	B0
1	1	C3	C2	C1	C0

19.5   Variable frequency divider
Design a circuit that will select (multiplex) a specified frequency to be output. A 200 KHz clock signal will be divided into a number of different frequencies by a 74LS393 counter chip. The desired frequency will be selected by a 74LS151, one-of-eight multiplexer as indicated in the following function table. The frequency selection is determined by the state of S2 S1 S0. The various frequencies are produced by the 74LS393 connected as shown in the logic diagram below. Use a frequency counter to measure the signal frequency output by the multiplexer.

S2	S1	S0	Output frequency (KHz)
0	0	0	0
0	0	1	1.25
0	1	0	2.5
0	1	1	5.0
1	0	0	10.0
1	0	1	20.0
1	1	0	100.0
1	1	1	200.0

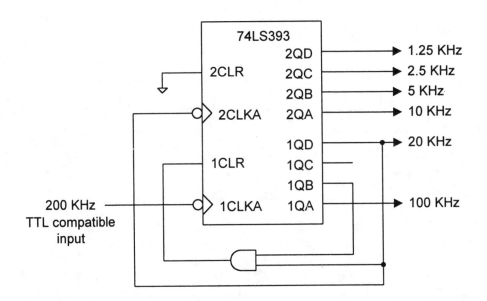

DEMULTIPLEXERS

Objective

- To apply demultiplexer circuits in digital system applications.

Suggested Parts					
74LS04	74LS10	74LS138	74LS151	74LS193	74LS393
GAL16V8					

Demultiplexer Circuits

A demultiplexer is also called a data distributor since it has several possible destinations for the input data to be sent to. The single output line that will receive the data is controlled by the specific select code applied to the demultiplexer. The 74LS138 (see Fig. 20-1) is an example MSI demultiplexer chip. This chip can serve the dual function of either a decoder or a demultiplexer. A decoder can be used as a demultiplexer by connecting the DMUX select lines to the data input lines (of the decoder) and applying the DMUX data to the enable input (of the decoder). One bit of input data may be sent to any one of eight possible output destinations with the 74LS138. Data that is input to one of the G2 enables will not be inverted by the DMUX chip. The other two enable pins may be utilized as enable controls.

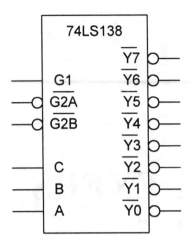

Fig. 20-1 74LS138, 8-channel demultiplexer chip

Laboratory Projects

Design demultiplexer circuits for the following applications. Breadboard and verify your designs.

20.1   Output line selection using a DMUX
Use the 74LS138 as a demultiplexer to drive any one of 4 possible lamps by a 1 Hz (approximate) clock signal. The specific lamp to receive the clock input will be selected by the control signals M and N. The selected output should be high when the clock input is high. Connect the clock signal to a lamp (Lamp 5) so that the output of the DMUX can be verified. Use the 74LS393 counter to sequence automatically the DMUX circuit by driving M with QD and N with QC. How many times does each lamp flash in turn with the counter controlling the sequence?

M	N	Lamp 1	Lamp 2	Lamp 3	Lamp 4
0	0	flashing	off	off	off
0	1	off	flashing	off	off
1	0	off	off	flashing	off
1	1	off	off	off	flashing

20.2   4-bit counter demultiplexer

Design and construct a 4-bit, 2-channel demultiplexer using a PLD so that a binary counter's output can be sent to one of 2 sets of lights controlled by the select input S. Use a 74LS393 binary counter to produce the 4-bit input data (QD QC QB QA) to the demultiplexer. Also include a gating control G on the demultiplexer circuit as indicated in the following function table.

G	S	A Lights				B Lights			
		A3	A2	A1	A0	B3	B2	B1	B0
0	0	QD	QC	QB	QA	0	0	0	0
0	1	0	0	0	0	QD	QC	QB	QA
1	X	0	0	0	0	0	0	0	0

20.3   Clock demultiplexer for 74LS193

Design a demultiplexer circuit to control the clocking of a 74LS193, 4-bit, up/down counter. The clock demultiplexer will have a control input to select the count direction and an enable control that can be used to stop the count sequence. The 74LS193 has two separate clock inputs, one to trigger the count up (UP) sequence and the other to trigger the count down (DOWN). Whichever clock input is not being used must be held high while the chip is clocked on the other input. The timing diagram below illustrates the operation of the clock demultiplexer. Hint: Use a single SSI chip for this DMUX circuit.

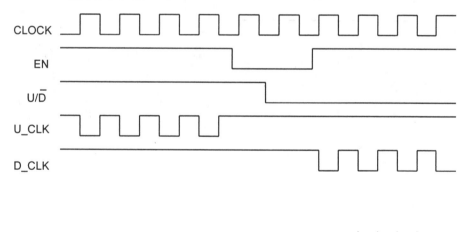

20.4   Serial data transmitter/receiver

Implement the data transmitter/receiver circuit given in the block diagram below using a 74LS151, a 74LS138, and the two counters in a 74LS393. Synchronize the two counters by resetting them at the same time. Clock the counters initially with a low frequency (approximately 1 Hz). What happens if the two counters are not synchronized together? Increase the clock frequency to approximately 1 KHz. What do you observe? What needs to be added to this circuit to make it more practical for serial data transmission?

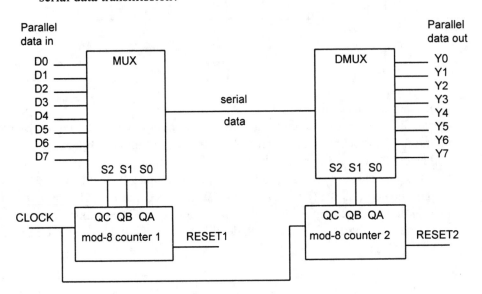

MAGNITUDE COMPARATORS

Objectives
- To design and implement comparator circuits using logic gates.
- To design and implement various circuit applications using comparator chips.

Suggested Parts			
74LS85	74LS191	74LS393	GAL16V8

Magnitude Comparators

Magnitude comparators are used to determine the magnitude relationships between two quantities. A typical comparator will indicate whether two input values are equivalent or, if not, which of the values is larger. Logic gates can be used to implement various types of comparator circuits, or MSI comparator chips such as the 74LS85, 4-bit comparator shown in Fig. 21-1 can be easily used in a variety of magnitude comparator applications. This comparator chip determines the magnitude relationship between the two binary values A3 A2 A1 A0 and B3 B2 B1 B0. The outputs produced are A>B, A=B, and A<B. The three cascade inputs allow multiple 74LS85 chips to be cascaded together to compare binary quantities that are more than 4-bits long. If the cascade feature is not used, the A=B input will normally be tied high and the other two cascade inputs (A>B and A<B) are tied low. An 8-bit magnitude comparator circuit

constructed with two 74LS85 chips is shown in Fig. 21-2. The two comparator chips are cascaded together by connecting the outputs from the comparator that is monitoring the least-significant 4 bits from each number to the cascade inputs on the comparator that is monitoring the most-significant 4 bits.

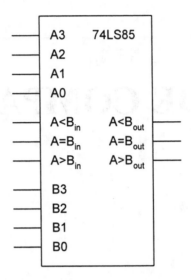

Fig. 21-1 74LS85, 4-bit magnitude comparator chip

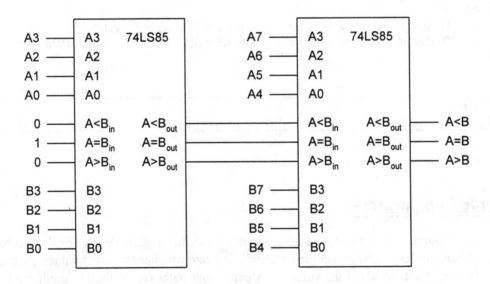

Fig. 21-2 8-bit magnitude comparator circuit using two 74LS85 chips

Laboratory Projects

Design logic circuits to implement each of the following comparator applications. Test and verify your designs.

21.1 Equal to or less than 9 comparator
Design a comparator circuit in which the single output will be high if the 4-bit data input value produced by a 74LS393 binary counter (QD QC QB QA) is equal to or less than 9. Use the 74LS85 comparator chip and any other necessary gates. Display both the counter and comparator outputs on lamps.

21.2 Data biasing circuit
Design a logic circuit to bias a 4-bit input value (I3 I2 I1 I0) to produce the 4-bit output (R3 R2 R1 R0) according to the following relationship. Use the 74LS85 and 74LS83 chips.

If:	Then:
I < 9	R = I + 1
I = 9	R = I
I > 9	R = I - 1

Where:
I = I3 I2 I1 I0
R = R3 R2 R1 R0

21.3 Programmable comparator
Design a comparator circuit that can compare a 4-bit data input value (I3 I2 I1 I0) with any one of 4 possible constants as indicated by the following table. S1 and S0 are control signals that determine which constant is selected for the B input to the comparator. Your circuit should provide the 3 output signals NE, GE, and LE. Use the 74LS85 comparator and any other necessary gates (or a PLD).

Controls		Constants			
S1	S0	B3	B2	B1	B0
0	0	0	1	1	1
0	1	1	0	0	0
1	0	1	0	0	1
1	1	1	0	1	0

Output signals needed:
NE = not equal
GE = greater than or equal to
LE = less than or equal to

21.4   Adjustable range detector

Design a logic circuit using a GAL16V8 that will compare a 5-bit input value (I4 I3 I2 I1 I0) to one of four specified input ranges selected by the controls R1 and R0. The comparator should have three outputs to indicate that the input value is either less than (LTR) or greater than (GTR) the range of values in the current window or within the range (RNG) of window values. Hint: Write the 3 logic expressions using the equality operator.

R1	R0	Selected range of window values:
0	0	16
0	1	15 thru 17
1	0	14 thru 18
1	1	13 thru 19

21.5   Variable modulus counter

Design a recycling counter circuit that can have any selected count modulus from 2 through 16. The desired count modulus will be determined by a 4-bit input value (mod = N3 N2 N1 N0). Mod-16 will be selected by N3 N2 N1 N0 = 0 0 0 0. Use a 74LS393 binary counter chip, a 74LS85 magnitude comparator, and any other necessary gates.

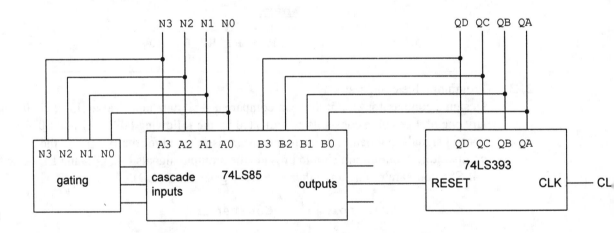

21.6   Number selector

Design a logic circuit that will output either the larger or smaller value as desired from two 4-bit inputs. The selection of the larger or smaller value is specified by the control signal F. Use a 74LS85 magnitude comparator chip to determine which 4-bit input is larger (or smaller) and a GAL16V8 to select the specified value. The PLD will function as a multiplexer in which the data is selected by the combination of the control F and the output of the magnitude comparator.

F	Selector output:
0	smaller value of A or B
1	larger value of A or B

Note: when A = B, select either A or B
        A = A3 A2 A1 A0
        B = B3 B2 B1 B0
        Y = Y3 Y2 Y1 Y0

Hint:

F	A>B	Y	
0	0	A	(select A input)
0	1	B	(select B input)
1	0	B	(select B input)
1	1	A	(select A input)

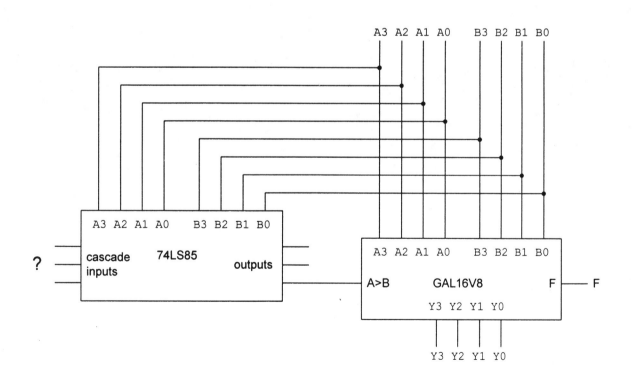

21.7 Variable self-stopping counter

Design a self-stopping counter circuit that can count up to any selected 4-bit binary value and then stop. The desired binary value at which to stop the counter will be determined by a 4-bit input (N3 N2 N1 N0). The counter should also have a manual, active-low RESTART control signal that will initialize the counter to 0 so that the counter can again cycle through its sequence. Use a 74LS191 binary counter chip and a 74LS85 magnitude comparator.

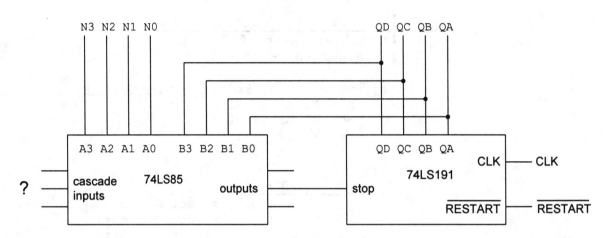

DIGITAL/ANALOG AND ANALOG/DIGITAL CONVERSION

Objectives

- To construct a digital-to-analog conversion (DAC) circuit using a standard commercial DAC IC chip.
- To construct an analog-to-digital conversion (ADC) circuit using a standard commercial ADC IC chip.

Suggested Parts	
AD557	ADC0804
Resistors:	3.3, 10 Kohm
Capacitor:	150 pf

Digital-to-Analog Conversion

Digital system outputs often must be interfaced to analog devices. One type of interfacing involves converting a digital data word into a representative analog signal (either a current or a voltage). Various digital-to-analog converter chips are available to perform this type of conversion. An example of a complete D/A converter in a single IC chip is the AD557 manufactured by Analog Devices. This 8-bit D/A converter is powered by a single 5-volt power supply and has a full-scale output of 2.55 volts. The resolution or incremental analog step size of the DAC is dependent on the number of digital input bits. The output voltage step size for this device is 10 mv. Fig. 22-1 shows how to use the AD557 to produce an analog output from 0 to 2.55 volts.

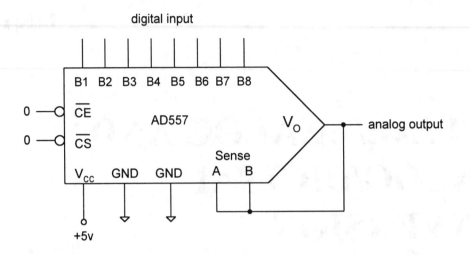

Fig. 22-1 Digital-to-analog conversion IC

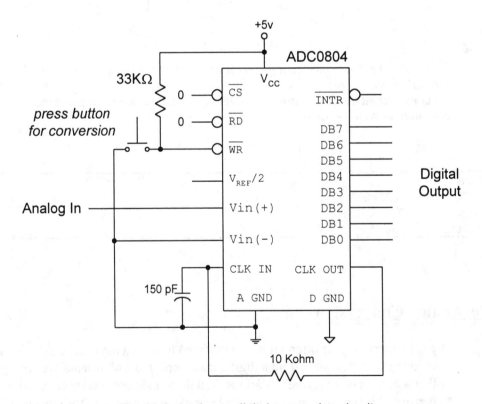

Fig. 22-2 Analog-to-digital conversion circuit

Analog-to-Digital Conversion

For a digital system to be able to process analog information, the analog signal must first be digitized or converted from analog to digital. The analog input signal is then represented by a digital value in the digital system. There are various analog-to-digital converter chips available to perform this type of conversion. The ADC0804 manufactured by National Semiconductor is an example of an 8-bit, successive approximation converter contained in a single IC chip. This device will handle an analog input range from 0 to 5 volts and is powered by a single 5-volt power supply. The quantization error of an ADC is dependent on the number of digital output bits. With a resolution of 8 bits and a full-scale output of 5 volts, the ADC0804 will have a quantization error of approximately 20 mv. Fig. 22-2 shows how to connect the ADC0804 so that it will digitize the analog input each time the pushbutton is pressed.

The ADC0804 ADC chip can also be operated in free-running mode by connecting (see Fig. 22-3) the end-of-conversion output signal INTR (interrupt) to the start-conversion input signal WR (write). It may be necessary to <u>momentarily</u> short the WR pin to ground in order to initially start the conversion process, but the ADC should then operate continuously. $V_{in(-)}$ and $V_{REF}/2$ can be changed to modify the analog input signal's span or range and step size. The analog differential input voltage ($V_{in(+)}$ - $V_{in(-)}$) is the actual signal that is converted to a representative digital output value. Therefore, the analog input voltage that produces an output of 00000000 can be adjusted above ground by applying a DC voltage to the $V_{in(-)}$ pin. The input voltage span can also be adjusted by applying a DC voltage to the $V_{REF}/2$ pin. The analog input voltage span will be equal to two times the voltage on the $V_{REF}/2$ pin.

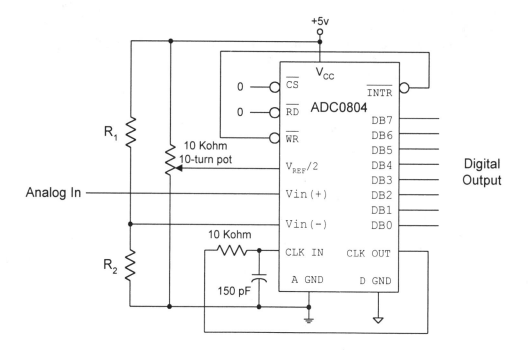

Fig. 22-3 Free-running analog-to-digital conversion circuit with span adjust

Laboratory Projects

22.1   Digital-to-analog converter
Construct a D/A converter using an Analog Devices AD557 IC chip. Determine a procedure to test the operation of the DAC. Graph the theoretical and test results for the DAC.

22.2   Analog-to-digital converter
Construct an A/D converter using a National ADC0804 IC chip. Determine a procedure to test the operation of the ADC. Graph the theoretical and test results for the ADC.

22.3   Free-running ADC with span adjust
Design a free-running A/D converter (using a National ADC0804) that produces a digital output of 00000000 for an analog input voltage of 0.5v and has a resolution of 15mv. Test your design.

22.4   Digital voltmeter
Construct a simple 0-5v DC digital voltmeter using the ADC0804 and a GAL16V8 to convert the binary output into BCD to be displayed on two 7-segment displays. The PLD will be programmed to convert a 6-bit binary number to a 2-digit BCD number. See diagram in Fig. 22-4. Note that the 2 least significant A/D output bits are left unconnected (using only 6 bits of resolution). Calibrate the digital voltmeter by applying 5v to the Analog In (measured with a DVM) and adjusting the 10-turn potentiometer until the 7-segment display reads 5.0 (the decimal point is assumed to be in the middle). Test the accuracy of the digital voltmeter with several different values for Analog In. Hint: See Unit 17 for the binary-to-BCD converter.

22.5   Digitized sine-wave generator
Construct a digitized sine-wave generator using an AD557 D/A converter, a GAL16V8, and a mod-24 binary counter. The sine-wave output will be produced by the D/A using data generated by the PLD. See diagram in Fig. 22-5. A mod-24 counter will provide 24 counts per cycle of the sine-wave. How many degrees are there per count? Calculate the D/A digital data needed for each step (count) to generate the desired analog output. The amplitude of the sine-wave should be 1.0 volt and the DC offset should be 1.5 volt. The formula for the instantaneous output voltage is:

$$v_{out} = 1.5 + 1.0 \sin \theta$$

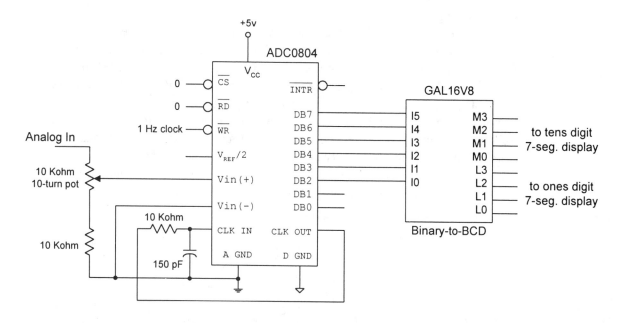

Fig. 22-4 Digital voltmeter diagram

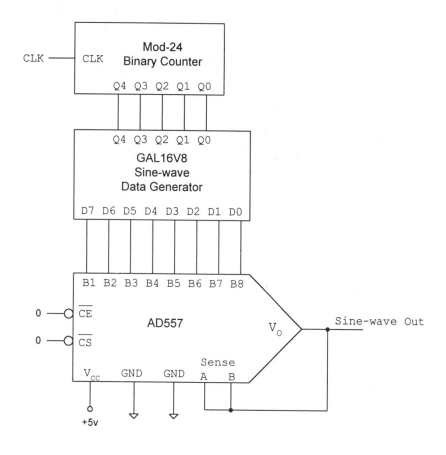

Fig. 22-5 Digitized sine-wave generator diagram

22.6   Reconstructing a digitized signal
Use the A/D converter to digitize a sine wave input signal and then reconvert the representative digital signal into an analog signal with the DAC. Connect the input of the D/A converter to the output of the A/D converter. Make sure the MSB out is connected to the MSB in. Adjust the input sine wave to 2.5 V_{p-p} with a DC offset of 1.25 volts (i.e., the sine wave goes from 0 to 2.5v). Adjust the $V_{REF}/2$ voltage using the potentiometer so that the analog input span on the ADC is 0 to 2.5v. Use an oscilloscope to compare the analog input and output waveforms. Set the analog input signal frequency to approximately 200 Hz and observe the resultant analog output signal. Sketch the input and output waveforms. Increase the input signal frequency and note the effect on the resultant output signal.

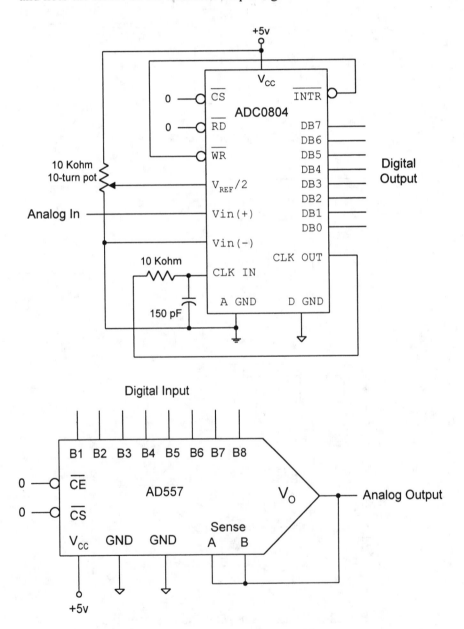

MEMORY SYSTEMS

Objectives

- To construct a random access memory (RAM) circuit using a semiconductor RAM chip.
- To read and write data in a RAM memory.
- To combine multiple RAM chips to expand word size and/or memory capacity.

Suggested Parts			
74LS04	74LS244	74LS393	2114

Memory Devices and Systems

Memory devices are used in digital systems to store digital information. A specific storage location in memory is identified with a unique binary value called an address. Accessing data stored at a specified address is referred to as a read operation. Storing new data at a specified address is referred to as a write operation.

Random access memory (RAM) is the term used frequently for memory that can be read from or written into with equal ease. Ordinary semiconductor RAM devices are said to be volatile because the information stored in the device is lost if the electrical power for the memory is removed. Semiconductor RAM may be either static, which does not need the stored data to be periodically refreshed, or dynamic, which does require the data to be periodically rewritten into the memory cells. Static RAM memory chips have the following types of pins: address pins that are used to select a specific memory location, data pins that are used to input the data into or output the data from the addressed memory location, a chip select pin to enable a specific chip (or set of chips) in a memory system, and a Read/Write pin to control the chip's read or write function (read = 1 and write = 0). The block diagram for a static RAM chip containing 256 4-bit words is shown in Fig. 23-1.

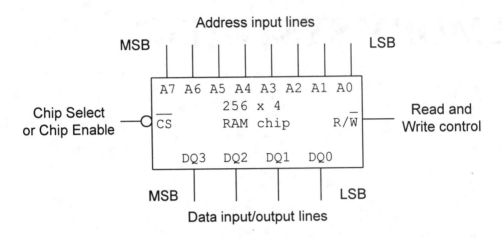

Fig. 23-1 Block diagram for a typical static RAM chip

Read only memory (ROM) often refers to a broad category of memory devices that normally have their data permanently stored in them while they are used in a digital system. There are various subtypes of ROM devices, some of which can have the stored data erased and replaced with new information.

Many different memory devices (with different part numbers) are available that differ in word size and memory capacity. The memory chip's word size is the number of bits that are accessed simultaneously in the chip with a given address. The memory capacity for a chip is the total number of words (2^n, where n = number of address bits) that can be addressed on the chip. Several memory chips can be interconnected to expand the total system memory. This expansion can be in word size, total number of addressable words, or both.

Tristate Bus Drivers

In addition to high and low output levels, a tristate output device also has a high impedance output condition. This type of output structure is normally used to connect several possible sources of digital information to a common bus, such as a data bus in a memory system. Only one source of information will be enabled at a time and allowed to place data on the bus. Most memory chips today have built-in tristate output buffers for the data pins that are enabled when the chip is selected and a read operation is performed. The source of data for a write operation must also be tristated from the data bus when a write operation is not being performed. The output for a tristated device will be a high impedance (acts like an open circuit) when the device is disabled. Separate tristate buffer or bus driver chips are available to accomplish this task. The 74LS244 shown in Fig. 23-2 is an example of a tristate buffer. It contains a total of eight buffers in one chip. The buffers are arranged in two sets of four buffers each. Each set of buffers has a common active-low tristate enable.

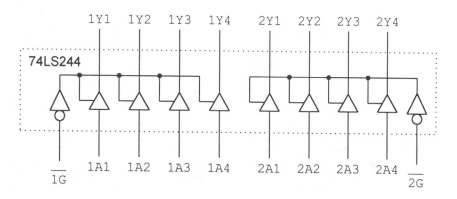

Fig. 23-2 74LS244, octal buffer with tristate output

Example 23-1

Design a 256-word RAM memory system in which each word is 8 bits. Use 1024-bit RAM memory chips that are arranged as 256 X 4 memory devices.

It will be necessary to use two of these 256 X 4 memory chips in order to double the word size for the specified memory system. The schematic for this 256 X 8 memory is shown in Fig. 23-3. The chips are permanently enabled with a 0 on the common (active low) chip select line. Both chips receive the same address information to access a 4-bit word in each. The two 4-bit words from each chip are used to double the word size for the memory system. RAM 0 provides the least significant 4 bits while RAM 1 provides the most significant 4 bits of the 8-bit data word. The octal tristate buffer chip (74LS244) is necessary so that external data (from some data source such as switches) can be stored in the memory system during a memory write cycle. During a memory write, the data from the switches is placed on the data bus and then is written into the addressed memory location. During a memory read cycle, the tristate buffers are disabled so that the data placed on the data bus comes from the memory chips.

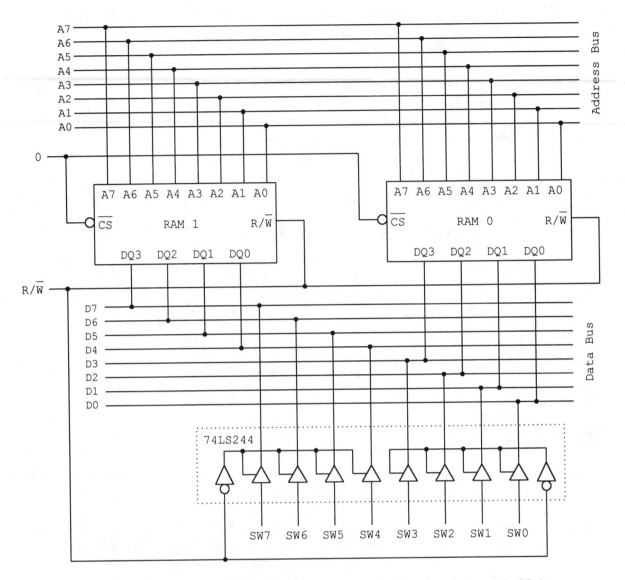

Fig. 23-3 Expanding word size for 256 x 8 RAM memory system in example 23-1

Example 23-2

Design a 512-word RAM memory system in which each word is 4-bits. Use 1024-bit RAM memory chips that are arranged as 256 X 4 memory devices.

It will be necessary to use two of these 256 X 4 RAM memory chips in order to double the word capacity for the specified memory system. The schematic for this 512 X 4 memory is shown in Fig. 23-4. Another address line (A8) will be necessary to double the word capacity that is provided by only one 256 X 4 chip. The A8 address line is decoded using an inverter so that only one chip is selected at a time. The other eight address lines are connected to each of the chip's address pins. If A8 is low, then RAM 0 will be accessed, or if A8 is high, then RAM 1 will be accessed. The corresponding data pins on each chip are tied to the four data bus lines. The tristate data output for each of the RAM chips will be enabled when the CS pin is low. The decoding gate (inverter) will allow only one of the chips to be enabled at a time. Only one-half (either half) of the octal tristate buffer chip will be needed since the data bus is only 4 bits wide in this memory system.

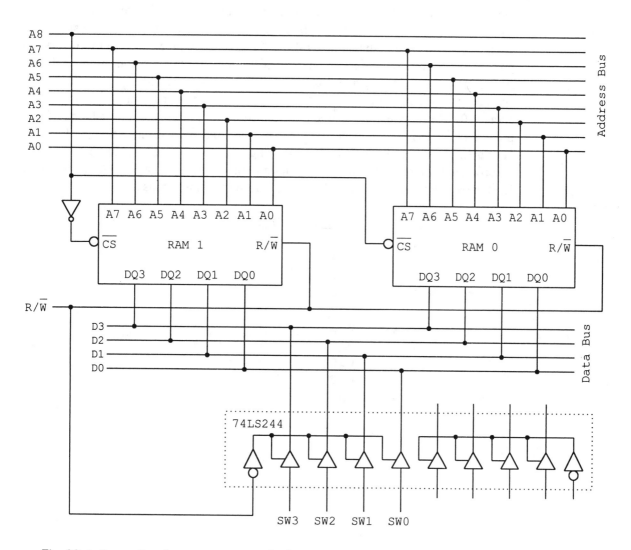

Fig. 23-4 Expanding the memory capacity for a 512 x 4 RAM memory system in example 23-2

Laboratory Projects

23.1 RAM memory chip

Test a 2114 (1024 X 4) memory chip by writing 4-bit data words to some selected memory locations and then verifying the data storage by reading the same selected addresses. Use a 74LS244 as the data bus buffer (enabled only during a write operation) for a set of 4 logic switches that provide the data to be stored in the RAM memory. Suggestion: Use a manually clocked, 8-bit binary counter (for example, cascade the two 4-bit counters in a 74LS393 together) connected to A_0 through A_7 to provide the address bus information. Connect the remaining two address pins (A8 and A9) to ground so that only one-fourth of the 2114 is being tested. Connect the address counter reset to a pushbutton so that you can easily start your memory system at address 0. This counter arrangement for the address bus information will make the project a little easier to manage. Monitor the data and address buses with LEDs. Test your memory system by writing the following sample hexadecimal data in the specified hex address (A9...A0) locations and then reading it back from the memory. Remember A9 = A8 = 0. Also check to see if the memory chips are volatile.

Address	Data
000	5
004	C
005	0
008	F
00B	6
00E	9
010	A
015	2
01F	1
020	7

23.2 Memory word-size expansion

Use two 2114 chips to design a 1024 X 8 memory system. Construct and test your design. Test your memory system by writing the following sample hexadecimal data in the specified hex address locations and then reading it back from the memory.

Address	Data
000	23
004	45
005	67
008	89
00B	AB
00E	CD
010	EF
015	01
01F	CC
020	33

23.3 Memory capacity expansion

Use two 2114 chips to design a 2048 X 4 memory system. Construct and test your design. Test your memory system by writing the following sample hexadecimal data in the specified hex address (A10...A0) locations and then reading it back from the memory.

Address	Data
000	1
400	2
007	3
407	4
00A	5
40A	6
010	7
410	8
017	9
417	A
01A	B
41A	C
020	D
420	E

LOGIC CIRCUIT TESTER

An inexpensive digital testing system having many of the capabilities of a commercially available system can be constructed with parts listed in the equipment list. The basic logic circuit tester described here may be either temporarily breadboarded as needed or mounted more permanently on some type of circuit board.

Lamp Monitor

A light emitting diode (LED) can be easily used as a lamp monitor for digital signals. A driver circuit (see Fig. A-1) using an inverter from a 74HCT05 can be used to turn on the LED without loading down the monitored gate's output.

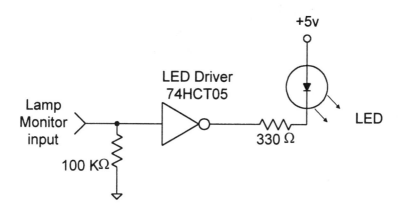

Fig. A-1 LED lamp monitor circuit

Logic Switches

An SPST toggle switch can be conveniently used for logic inputs in digital circuit testing. Fig. A-2 shows how to wire a simple logic switch.

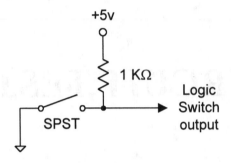

Fig. A-2 Logic switch circuit

Pushbutton

A pushbutton or momentary switch can be easily debounced, thereby making it suitable for use in digital circuits, using a simple NAND latch as shown in Fig. A-3.

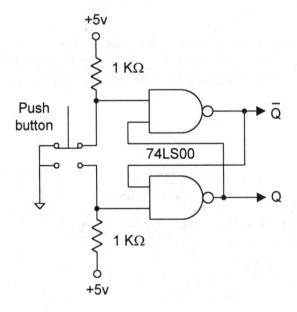

Fig. A-3 Debounced switch circuit

Clock

A 555 timer chip can be conveniently used to produce a TTL compatible square wave for clocking digital circuits. The frequency may be varied by changing the timing components R_A, R_B, and C. Fig. A-4 shows the clock circuit schematic and equations for determining the output frequency.

$$Frequency = 1.44/(R_A + 2R_B)C$$

$$duty\ cycle = [(R_A + R_B)/(R_A + 2R_B)] \times 100\%$$

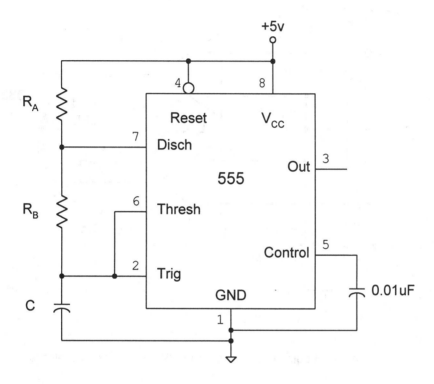

Fig. A-4 Clock circuit using a 555 timer

Logic Probe

A TTL voltage-compatible logic probe can be easily constructed using two analog comparators as shown in Fig. A-5. The red LED will light for input voltages above approximately 2v and the green LED should light for input voltages below 0.8v.

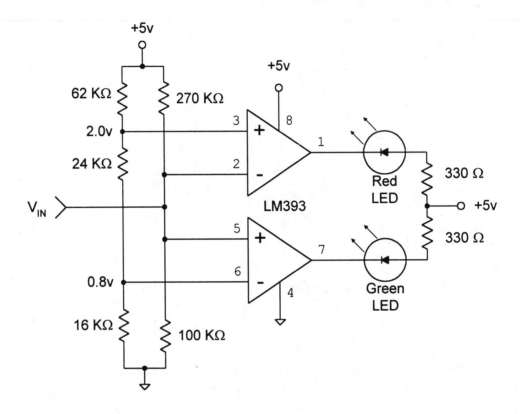

Fig. A-5 Logic probe using an LM393 analog comparator

PLD DESIGN VERIFICATION USING THE CUPLWIN SIMULATOR

CUPLWIN also can perform logic simulation of the PLD design. There are actually two applications for this simulator. It can be used to perform a functional simulation on the logic of a PLD design to determine if the design has been defined correctly. It can also be used to create test vectors that can be downloaded to many PLD programmers to perform a functional test on the actual PLD device after it has been programmed with the design.

An input file (filename.SI) that contains the desired test specifications is created using a text editor. This file contains a sequential list of test vectors that includes the input pin stimuli and the expected output pin results. For logic simulation of the design, these test vectors are compared with the actual results as determined by the logic definition given in the CUPL source file. The calculated output values for the design are contained in a file called an absolute file (filename.ABS). The absolute file must be created before you can simulate the design. The absolute file can be generated by CUPL when the design is compiled. This is accomplished by selecting the Option menu and then choosing Compiler Options. Click on Output file in the Compiler Options dialog box, select *absolute* in the Output Format dialog box, and then click OK. The design will be automatically simulated after (a successful) compilation if *simulate* is selected under Miscellaneous in the Compiler Options dialog box.

A simulation output listing file (filename.SO) can be generated by the simulator. This file contains the results of the simulation. The simulation output file will be generated by selecting the Option menu and then choosing Simulator Options. Select *Listing File* in the Simulate Options dialog box and then click OK. The test vectors are numbered in the listing file (filename.SO) and the inputs and outputs are listed in a tabular form in the order specified by the filename.SI input file. Any output tests that failed in the simulation are flagged and the actual output value for the design is listed. Each output that fails is marked and the user-expected result is listed along with an appropriate error message.

Test vectors to test the programmed device inserted in the programmer can be added to the JEDEC programming file (filename.JED). The test vectors can be stored in the filename.jed file by selecting *Append Vectors* in the Simulate Options dialog box.

After the PLD design has been compiled (with the *absolute* option) and if the simulation input file has already been created, you can just perform the simulation by selecting the Run menu and then choosing Device Specific Simulate. If any changes are made to the design file (filename.PLD) it must be recompiled and a new absolute file created before it can be simulated.

Test Vector Values	
Input Values	**Description**
0	drive input low
1	drive input high
X	input high or low
C	drive clock input (low-high-low)
K	drive clock input (high-low-high)
P	preload internal registers
Output Values	**Description**
L	test output for low
H	test output for high
X	output high or low
Z	test output for high impedance
N	output not tested
*	simulator determines output value

Table B-1 Test vector values in CSIM

The simulation input (test specification) file (filename.SI) normally contains the same header information as the CUPL logic description file (filename.PLD). If any header information is different, a warning message will suggest that the status of the logic equations could be inconsistent with the current test vectors in the test specification file. The easiest way to create the simulation input file is to copy the logic description file to the filename.SI file and then delete everything except the header information from the simulation input file. Comments (/* · · · */) may be placed anywhere within the test specification file. The simulator keyword **ORDER** is used to list the variables to be used in the simulation table and to define the order in which you wish them to be displayed. In the ORDER statement, a colon is placed after the word ORDER, then each input/output variable (separated by commas) is listed, and the list is terminated by a semicolon. Column spaces can be inserted in the simulation data table with the order statement by using the percent symbol (%) followed by the number of (decimal) spaces desired. The test vector table is prefixed by the simulator keyword **VECTORS**, which is followed by a colon. Each sequential test vector must be specified on a single line. The possible input and output values for each test vector are shown in Table B-1. Notice that input variables use 0 and 1 and that output variables use L and H for logic levels.

Example B-1

Create a test specification file and simulate the design in example 6-2.

The simulation input file is shown in Fig. B-1. The desired order for the inputs and outputs is specified in the ORDER statement. Two blank spaces are specified between the last input bit (I0) and the first output bit (N7). Also a single blank space is specified between the upper 4 bits and the lower 4 bits of the output. The spacing was done to improve the readability of the vector table. The VECTORS statement is followed by the vector table which lists the input values and the expected output results. The results of this simulation are shown in Fig. B-2. No errors were detected in the simulation (since there were no error messages).

Example B-2

Modify the alternate design for example 6-1 (see Fig. 6-6) so that the effect of errors in logic equations may be investigated. The same logic functions are desired but the equations are incorrectly defined. Determine the effect on the simulation results. Change the equations for GT9 and LT4 to the following incorrect descriptions:

```
GT9 = INPUTS:[9..F];
LT4 = INPUTS:[0,1,2,3,4];
```

The correct simulation input file is shown in Fig. B-3. Simulating the design with the errors in the logic equations produced the results shown in Fig. B-4.

Example B-3

Create a test specification file and simulate the design in example 13-5.

The simulation input file is shown in Fig. B-5. The test vectors for this simulation were arbitrarily chosen, but an attempt was made to test all functions of this circuit design. List notation was used for appropriate variables in the ORDER statement. Each test vector is commented to assist in interpretation of the simulation. The input variable "C" was used to provide a positive pulse to clock the sequential PLD circuit.

The input value "X" was used several times when the input logic did not matter in the test vector. three lines of documentation messages for the simulator output file. The simulator directive **$msg** is used to place documentation messages in the simulator output file. All simulator directives begin with a "$" and each directive ends with a semicolon. The desired text string in the message is placed in double quotes. The simulator directive **$repeat** causes the vector <u>immediately following</u> to be repeated a specified (in decimal) number of times. The output test values during the repeat directive are not specified since they will be different values as the counter is clocked repeatedly. Instead an asterisk causes simulator to supply the output test value. The resultant simulation output file is shown in Fig. B-6.

```
Name       SQUARE;
Partno     L105-2;
Date       02/04/91;
Revision   03;
Designer   Greg Moss;
Company    Digi-Lab, Inc.;
Assembly   SQUARE NUMBER GENERATOR;
Location   U210;
Device     G16V8A;
Format     j;

/************************************************/
/* Generates the square of an input value using */
/* the truth table design entry technique.      */
/************************************************/

ORDER:  I3,I2,I1,I0,%2,N7,N6,N5,N4,%1,N3,N2,N1,N0;

VECTORS:
0000  LLLL LLLL
0001  LLLL LLLH
0010  LLLL LHLL
0011  LLLL HLLH
0100  LLLH LLLL
0101  LLLH HLLH
0110  LLHL LHLL
0111  LLHH LLLH
1000  LHLL LLLL
1001  LHLH LLLH
1010  LHHL LHLL
1011  LHHH HLLH
1100  HLLH LLLL
1101  HLHL HLLH
1110  HHLL LHLL
1111  HHHL LLLH
```

Fig. B-1 Computer listing of SQUARE.SI for example B-1

```
CSIM(WM): CUPL Simulation Program
Version 4.7a Serial# MW-66999998
Copyright (c) 1983, 1996 Logical Devices, Inc.
CREATED Thu Dec 19 14:23:36 1996

LISTING FOR SIMULATION FILE: SQUARE.si

    1: Name        SQUARE;
    2: Partno      L105-2;
    3: Date        02/04/91;
    4: Revision    03;
    5: Designer    Greg Moss;
    6: Company     Digi-Lab, Inc.;
    7: Assembly    SQUARE NUMBER GENERATOR;
    8: Location    U210;
    9: Device      G16V8A;
   10: Format      j;
   11:
   12: /**********************************************/
   13: /* Generates the square of an input value using */
   14: /* the truth table design entry technique.      */
   15: /**********************************************/
   16:
   17: FIELD data_in = [I3,I2,I1,I0];
   18: FIELD square = [N7,N6,N5,N4,N3,N2,N1,N0];
   19:
   20: ORDER:  I3,I2,I1,I0,%2,N7,N6,N5,N4,%1,N3,N2,N1,N0;
   21:

=========================
     IIII  NNNN NNNN
     3210  7654 3210
=========================
0001: 0000  LLLL LLLL
0002: 0001  LLLL LLLH
0003: 0010  LLLL LHLL
0004: 0011  LLLL HLLH
0005: 0100  LLLH LLLL
0006: 0101  LLLH HLLH
0007: 0110  LLHL LHLL
0008: 0111  LLHH LLLH
0009: 1000  LHLL LLLL
0010: 1001  LHLH LLLH
0011: 1010  LHHL LHLL
0012: 1011  LHHH HLLH
0013: 1100  HLLH LLLL
0014: 1101  HLHL HLLH
0015: 1110  HHLL LHLL
0016: 1111  HHHL LLLH
```

Fig. B-2 Computer listing of SQUARE.SO for example B-1

```
Name        LAB-ERR;
Partno      L105-1;
Date        12/17/96;
Revision    01;
Designer    Greg Moss;
Company     Digi-Lab, Inc;
Assembly    Example Board;
Location    U101;
Device      G16V8A;
Format      j;

    /*****************************************/
    /* ERROR in example 6-1 circuit design   */
    /* to illustrate CSIM results            */
    /*****************************************/

ORDER:  INPUTS,%3,GT9,%1,!LT4,%1,RNG,%1,!TEN;

VECTORS:

0000    L L L H
0001    L L L H
0010    L L L H
0011    L L L H
0100    L H L H
0101    L H L H
0110    L H L H
0111    L H H H
1000    L H H H
1001    L H H H
1010    H H H L
1011    H H L H
1100    H H L H
1101    H H L H
1110    H H L H
1111    H H L H
```

Fig. B-3 Computer listing of simulation input file for example B-2

```
CSIM(WM): CUPL Simulation Program
Version 4.7a Serial# MW-66999998
Copyright (c) 1983, 1996 Logical Devices, Inc.
CREATED Thu Dec 19 13:48:03 1996
LISTING FOR SIMULATION FILE: lab-err.si

   1: Name        LAB-ERR;
   2: Partno      L105-1;
   3: Date        12/17/96;
   4: Revision    01;
   5: Designer    Greg Moss;
   6: Company     Digi-Lab, Inc;
   7: Assembly    Example Board;
   8: Location    U101;
   9: Device      G16V8A;
  10: Format      j;
  11:
  12:      /*****************************************/
  13:      /* ERROR in example 6-1 circuit design   */
  14:      /* to illustrate CSIM results            */
  15:      /*****************************************/
  16:
  17: FIELD INPUTS = [D,C,B,A];
  18:
  19: ORDER:  INPUTS,%3,GT9,%1,!LT4,%1,RNG,%1,!TEN;
  20:


=======================
               !   !
             G L R T
             T T N E
     INPUTS  9 4 G N
=======================
0001: 0000   L L L H
0002: 0001   L L L H
0003: 0010   L L L H
0004: 0011   L L L H
0005: 0100   L L L H
               ^
[0019sa] user expected (H) for LT4

0006: 0101   L H L H
0007: 0110   L H L H
0008: 0111   L H H H
0009: 1000   L H H H
0010: 1001   H H H H
               ^
[0019sa] user expected (L) for GT9

0011: 1010   H H H L
0012: 1011   H H L H
0013: 1100   H H L H
0014: 1101   H H L H
0015: 1110   H H L H
0016: 1111   H H L H
```

Fig. B-4 Computer listing of simulation output file for example B-2

```
Name       LOADCNTR;              Partno    L155-17E;
Date       12/17/96;              Revision  01;
Designer   G. Moss;               Company   Digi-Lab;
Assembly   Controller;            Location  U922;
Device     G16V8A;                Format    j;

/*********************************************************/
/* Mod-16 binary up/down counter with parallel load      */
/*********************************************************/

ORDER:  clk, %2, M2..0, %2, D3..0, %2, !oe, %4, Q3..0;

VECTORS:
$msg "initialize counter by resetting";
           C   11X   XXXX   0   LLLL      /* reset counter          */
           C   00X   XXXX   0   LLLL      /* hold count             */
           0   XXX   XXXX   0   LLLL      /* no clock               */
           C   100   XXXX   0   LLLH      /* count up               */
           C   100   XXXX   0   LLHL      /* count up               */
           C   100   XXXX   0   LLHH      /* count up               */
           C   100   XXXX   0   LHLL      /* count up               */
$repeat 4;
           C   100   XXXX   0   ****      /* count up 4 cycles      */
           0   100   XXXX   0   HLLL      /* no clock               */
           C   100   XXXX   0   HLLH      /* count up               */
           C   000   XXXX   0   HLLH      /* hold count             */
$repeat 6;
           C   100   XXXX   0   ****      /* count up 5 cycles      */
           C   001   XXXX   0   HHHH      /* hold count             */
           C   100   XXXX   0   LLLL      /* count up - recycled    */
$msg "try counting down";
           C   101   XXXX   0   HHHH      /* count down             */
           C   101   XXXX   0   HHHL      /* count down             */
$repeat 8;
           C   101   XXXX   0   ****      /* count down 8 cycles    */
           0   101   XXXX   0   LHHL      /* no clock               */
           C   00X   XXXX   0   LHHL      /* hold count             */
$msg "try loading a number";
           C   010   1010   0   HLHL      /* load 1010              */
           C   101   XXXX   1   ZZZZ      /* count down,tri-state   */
           C   100   XXXX   0   HLHL      /* count up               */
$repeat 16;
           C   101   XXXX   0   ****      /* count down 16 cycles   */
           C   000   XXXX   0   HLHL      /* hold count             */
           C   011   0101   0   LHLH      /* load 0101              */
           C   01X   0000   0   LLLL      /* load 0000              */
$repeat 10;
           C   100   0000   0   ****      /* count up 10 cycles     */
$repeat 9;
           C   101   0000   0   ****      /* count down 9 cycles    */
           0   101   1111   0   LLLH      /* no clock               */
           C   111   1111   0   LLLL      /* reset counter          */
```

Fig. B-5 Computer listing of simulation input file for example B-3

```
CSIM(WM): CUPL Simulation Program
Version 4.7a Serial# MW-66999998
Copyright (c) 1983, 1996 Logical Devices, Inc.
CREATED Thu Jan 16 15:18:11 1997

LISTING FOR SIMULATION FILE: LOADCNTR.si

   1: Name       LOADCNTR;        Partno    L155-17E;
   2: Date       01/15/97;        Revision  01;
   3: Designer   G. Moss;         Company   Digi-Lab;
   4: Assembly   Controller;      Location  U922;
   5: Device     G16V8A;          Format    j;
   6:
   7: /*****************************************************/
   8: /* Mod-16 binary up/down counter with parallel load  */
   9: /*****************************************************/
  10:
  11: FIELD counter = [Q3,Q2,Q1,Q0];
  12: FIELD data = [D3,D2,D1,D0];
  13: FIELD mode = [M2,M1,M0];
  14:
  15: ORDER:  clk, %2, M2..0, %2, D3..0, %2, !oe, %4, Q3..0;
  16:


=================================
      c               !
      l   MMM  DDDD   o    QQQQ
      k   210  3210   e    3210
=================================
initialize counter by resetting
0001: C   11X   XXXX   0    LLLL
0002: C   00X   XXXX   0    LLLL
0003: 0   XXX   XXXX   0    LLLL
0004: C   100   XXXX   0    LLLH
0005: C   100   XXXX   0    LLHL
0006: C   100   XXXX   0    LLHH
0007: C   100   XXXX   0    LHLL
0008: C   100   XXXX   0    LHLH
0009: C   100   XXXX   0    LHHL
0010: C   100   XXXX   0    LHHH
0011: C   100   XXXX   0    HLLL
0012: 0   100   XXXX   0    HLLL
0013: C   100   XXXX   0    HLLH
0014: C   000   XXXX   0    HLLH
0015: C   100   XXXX   0    HLHL
0016: C   100   XXXX   0    HLHH
0017: C   100   XXXX   0    HHLL
0018: C   100   XXXX   0    HHLH
0019: C   100   XXXX   0    HHHL
0020: C   100   XXXX   0    HHHH
0021: C   001   XXXX   0    HHHH
0022: C   100   XXXX   0    LLLL
try counting down
0023: C   101   XXXX   0    HHHH
0024: C   101   XXXX   0    HHHL
0025: C   101   XXXX   0    HHLH
0026: C   101   XXXX   0    HHLL
0027: C   101   XXXX   0    HLHH
```

```
0028: C  101  XXXX  0   HLHL
0029: C  101  XXXX  0   HLLH
0030: C  101  XXXX  0   HLLL
0031: C  101  XXXX  0   LHHH
0032: C  101  XXXX  0   LHHL
0033: 0  101  XXXX  0   LHHL
0034: C  00X  XXXX  0   LHHL
try loading a number
0035: C  010  1010  0   HLHL
0036: C  101  XXXX  1   ZZZZ
0037: C  100  XXXX  0   HLHL
0038: C  101  XXXX  0   HLLH
0039: C  101  XXXX  0   HLLL
0040: C  101  XXXX  0   LHHH
0041: C  101  XXXX  0   LHHL
0042: C  101  XXXX  0   LHLH
0043: C  101  XXXX  0   LHLL
0044: C  101  XXXX  0   LLHH
0045: C  101  XXXX  0   LLHL
0046: C  101  XXXX  0   LLLH
0047: C  101  XXXX  0   LLLL
0048: C  101  XXXX  0   HHHH
0049: C  101  XXXX  0   HHHL
0050: C  101  XXXX  0   HHLH
0051: C  101  XXXX  0   HHLL
0052: C  101  XXXX  0   HLHH
0053: C  101  XXXX  0   HLHL
0054: C  000  XXXX  0   HLHL
0055: C  011  0101  0   LHLH
0056: C  01X  0000  0   LLLL
0057: C  100  0000  0   LLLH
0058: C  100  0000  0   LLHL
0059: C  100  0000  0   LLHH
0060: C  100  0000  0   LHLL
0061: C  100  0000  0   LHLH
0062: C  100  0000  0   LHHL
0063: C  100  0000  0   LHHH
0064: C  100  0000  0   HLLL
0065: C  100  0000  0   HLLH
0066: C  100  0000  0   HLHL
0067: C  101  0000  0   HLLH
0068: C  101  0000  0   HLLL
0069: C  101  0000  0   LHHH
0070: C  101  0000  0   LHHL
0071: C  101  0000  0   LHLH
0072: C  101  0000  0   LHLL
0073: C  101  0000  0   LLHH
0074: C  101  0000  0   LLHL
0075: C  101  0000  0   LLLH
0076: 0  101  1111  0   LLLH
0077: C  111  1111  0   LLLL
```

Fig. B-6 Computer listing of simulation output file for example B-3

APPENDIX C

MANUFACTURERS' DATA SHEETS

Courtesy of:
    Analog Devices, Inc.
    Lattice Semiconductor Corp.
    Texas Instruments Inc.

- **Package Options Include Plastic "Small Outline" Packages, Ceramic Chip Carriers and Flat Packages, and Plastic and Ceramic DIPs**

- **Dependable Texas Instruments Quality and Reliability**

description

These devices contain four independent 2-input-NAND gates.

The SN5400, SN54LS00, and SN54S00 are characterized for operation over the full military temperature range of −55 °C to 125 °C. The SN7400, SN74LS00, and SN74S00 are characterized for operation from 0 °C to 70 °C.

FUNCTION TABLE (each gate)

INPUTS		OUTPUT
A	**B**	**Y**
H	H	L
L	X	H
X	L	H

logic symbol†

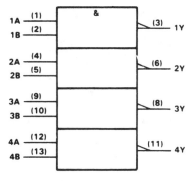

† This symbol is in accordance with ANSI/IEEE Std. 91-1984 and IEC Publication 617-12.

Pin numbers shown are for D, J, and N packages.

SN5400 . . . J PACKAGE
SN54LS00, SN54S00 . . . J OR W PACKAGE
SN7400 . . . N PACKAGE
SN74LS00, SN74S00 . . . D OR N PACKAGE
(TOP VIEW)

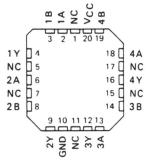

SN5400 . . . W PACKAGE
(TOP VIEW)

SN54LS00, SN54S00 . . . FK PACKAGE
(TOP VIEW)

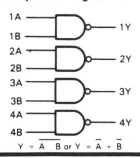

NC - No internal connection

logic diagram (positive logic)

$$Y = \overline{A \cdot B} \text{ or } Y = \overline{A} + \overline{B}$$

2

TTL Devices

TEXAS INSTRUMENTS
POST OFFICE BOX 655012 • DALLAS, TEXAS 75265

SN5400, SN54LS00, SN54S00,
SN7400, SN74LS00, SN74S00
QUADRUPLE 2-INPUT POSITIVE-NAND GATES

schematics (each gate)

'00

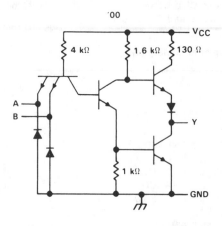

'LS00

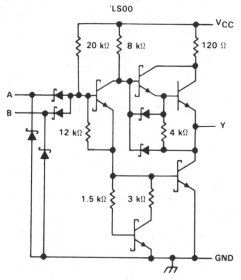

'S00

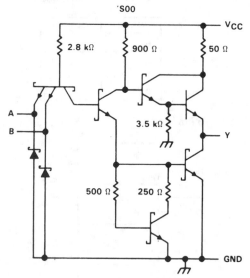

Resistor values shown are nominal.

absolute maximum ratings over operating free-air temperature range (unless otherwise noted)

Supply voltage, V_{CC} (see Note 1) . . . . 7 V
Input voltage: '00, 'S00 . . . . 5.5 V
'LS00 . . . . 7 V
Operating free-air temperature range: SN54' . . . . −55°C to 125°C
SN74' . . . . 0°C to 70°C
Storage temperature range . . . . −65°C to 150°C

NOTE 1: Voltage values are with respect to network ground terminal.

TEXAS
INSTRUMENTS

POST OFFICE BOX 655012 • DALLAS, TEXAS 75265

recommended operating conditions

		SN5400			SN7400			UNIT
		MIN	NOM	MAX	MIN	NOM	MAX	
V_{CC}	Supply voltage	4.5	5	5.5	4.75	5	5.25	V
V_{IH}	High-level input voltage	2			2			V
V_{IL}	Low-level input voltage			0.8			0.8	V
I_{OH}	High-level output current			−0.4			−0.4	mA
I_{OL}	Low-level output current			16			16	mA
T_A	Operating free-air temperature	−55		125	0		70	°C

electrical characteristics over recommended operating free-air temperature range (unless otherwise noted)

PARAMETER	TEST CONDITIONS†			SN5400			SN7400			UNIT
				MIN	TYP‡	MAX	MIN	TYP‡	MAX	
V_{IK}	V_{CC} = MIN,	I_I = −12 mA				−1.5			−1.5	V
V_{OH}	V_{CC} = MIN,	V_{IL} = 0.8 V,	I_{OH} = −0.4 mA	2.4	3.4		2.4	3.4		V
V_{OL}	V_{CC} = MIN,	V_{IH} = 2 V,	I_{OL} = 16 mA		0.2	0.4		0.2	0.4	V
I_I	V_{CC} = MAX,	V_I = 5.5 V				1			1	mA
I_{IH}	V_{CC} = MAX,	V_I = 2.4 V				40			40	µA
I_{IL}	V_{CC} = MAX,	V_I = 0.4 V				−1.6			−1.6	mA
I_{OS}§	V_{CC} = MAX			−20		−55	−18		−55	mA
I_{CCH}	V_{CC} = MAX,	V_I = 0 V			4	8		4	8	mA
I_{CCL}	V_{CC} = MAX,	V_I = 4.5 V			12	22		12	22	mA

† For conditions shown as MIN or MAX, use the appropriate value specified under recommended operating conditions.
‡ All typical values are at V_{CC} = 5 V, T_A = 25°C.
§ Not more than one output should be shorted at a time.

switching characteristics, V_{CC} = 5 V, T_A = 25°C (see note 2)

PARAMETER	FROM (INPUT)	TO (OUTPUT)	TEST CONDITIONS		MIN	TYP	MAX	UNIT
t_{PLH}	A or B	Y	R_L = 400 Ω,	C_L = 15 pF		11	22	ns
t_{PHL}						7	15	ns

NOTE 2: Load circuits and voltage waveforms are shown in Section 1.

2 TTL Devices

Texas Instruments
POST OFFICE BOX 655012 • DALLAS, TEXAS 75265

QUADRUPLE 2-INPUT POSITIVE-NAND GATES

recommended operating conditions

		SN54LS00			SN74LS00			UNIT
		MIN	NOM	MAX	MIN	NOM	MAX	
V_{CC}	Supply voltage	4.5	5	5.5	4.75	5	5.25	V
V_{IH}	High-level input voltage	2			2			V
V_{IL}	Low-level input voltage			0.7			0.8	V
I_{OH}	High-level output current			−0.4			−0.4	mA
I_{OL}	Low-level output current			4			8	mA
T_A	Operating free-air temperature	−55		125	0		70	°C

electrical characteristics over recommended operating free-air temperature range (unless otherwise noted)

PARAMETER	TEST CONDITIONS †			SN54LS00			SN74LS00			UNIT
			MIN	TYP‡	MAX	MIN	TYP‡	MAX	
V_{IK}	V_{CC} = MIN,	I_I = −18 mA			−1.5			−1.5	V
V_{OH}	V_{CC} = MIN,	V_{IL} = MAX, I_{OH} = −0.4 mA	2.5	3.4		2.7	3.4		V
V_{OL}	V_{CC} = MIN,	V_{IH} = 2 V, I_{OL} = 4 mA		0.25	0.4		0.25	0.4	V
	V_{CC} = MIN,	V_{IH} = 2 V, I_{OL} = 8 mA					0.35	0.5	
I_I	V_{CC} = MAX,	V_I = 7 V			0.1			0.1	mA
I_{IH}	V_{CC} = MAX,	V_I = 2.7 V			20			20	µA
I_{IL}	V_{CC} = MAX,	V_I = 0.4 V			−0.4			−0.4	mA
I_{OS} §	V_{CC} = MAX		−20		−100	−20		−100	mA
I_{CCH}	V_{CC} = MAX,	V_I = 0 V		0.8	1.6		0.8	1.6	mA
I_{CCL}	V_{CC} = MAX,	V_I = 4.5 V		2.4	4.4		2.4	4.4	mA

† For conditions shown as MIN or MAX, use the appropriate value specified under recommended operating conditions.
‡ All typical values are at V_{CC} = 5 V, T_A = 25°C
§ Not more than one output should be shorted at a time, and the duration of the short-circuit should not exceed one second.

switching characteristics, V_{CC} = 5 V, T_A = 25°C (see note 2)

PARAMETER	FROM (INPUT)	TO (OUTPUT)	TEST CONDITIONS		MIN	TYP	MAX	UNIT
t_{PLH}	A or B	Y	R_L = 2 kΩ,	C_L = 15 pF		9	15	ns
t_{PHL}						10	15	ns

NOTE 2: Load circuits and voltage waveforms are shown in Section 1.

2

TTL Devices

TEXAS
INSTRUMENTS
POST OFFICE BOX 655012 • DALLAS, TEXAS 75265

SN54HC00, SN74HC00
QUADRUPLE 2-INPUT POSITIVE-NAND GATES

D2684, DECEMBER 1982 – REVISED MARCH 1984

- **Package Options Include Plastic "Small Outline" Packages, Ceramic Chip Carriers, and Standard Plastic and Ceramic 300-mil DIPs**

- **Dependable Texas Instruments Quality and Reliability**

description

These devices contain four independent 2-input NAND gates. They perform the Boolean functions $Y = \overline{A \cdot B}$ or $Y = \overline{A} + \overline{B}$ in positive logic.

The SN54HC00 is characterized for operation over the full military temperature range of −55°C to 125°C. The SN74HC00 is characterized for operation from −40°C to 85°C.

FUNCTION TABLE (each gate)

INPUTS		OUTPUT
A	**B**	**Y**
H	H	L
L	X	H
X	L	H

logic symbol†

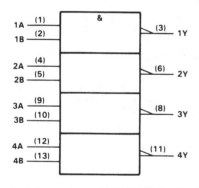

† This symbol is in accordance with ANSI/IEEE Std 91-1984 and IEC Publication 617-12.
Pin numbers shown are for D, J, or N packages.

SN54HC00 . . . J PACKAGE
SN74HC00 . . . D OR N PACKAGE
(TOP VIEW)

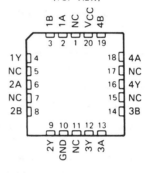

SN54HC00 . . . FK PACKAGE
(TOP VIEW)

NC – No internal connection

logic diagram (each gate)

HCMOS Devices

2

Copyright © 1982, Texas Instruments Incorporated

TEXAS
INSTRUMENTS
POST OFFICE BOX 655012 • DALLAS, TEXAS 75265

SN54HC00, SN74HC00
QUADRUPLE 2-INPUT POSITIVE-NAND GATES

absolute maximum ratings over operating free-air temperature range[†]

Supply voltage, V_{CC}	-0.5 V to 7 V
Input clamp current, I_{IK} ($V_I < 0$ or $V_I > V_{CC}$)	± 20 mA
Output clamp current, I_{OK} ($V_O < 0$ or $V_O > V_{CC}$)	± 20 mA
Continuous output current, I_O ($V_O = 0$ to V_{CC})	± 25 mA
Continuous current through V_{CC} or GND pins	± 50 mA
Lead temperature 1,6 mm (1/16 in) from case for 60 s: FK or J package	300 °C
Lead temperature 1,6 mm (1/16 in) from case for 10 s: D or N package	260 °C
Storage temperature range	-65 °C to 150 °C

[†] Stresses beyond those listed under "absolute maximum ratings" may cause permanent damage to the device. These are stress ratings only, and functional operation of the device at these or any other conditions beyond those indicated under "recommended operating conditions" is not implied. Exposure to absolute-maximum-rated conditions for extended periods may affect device reliability.

recommended operating conditions

			SN54HC00			SN74HC00			UNIT
			MIN	NOM	MAX	MIN	NOM	MAX	
V_{CC}	Supply voltage		2	5	6	2	5	6	V
V_{IH}	High-level input voltage	V_{CC} = 2 V	1.5			1.5			V
		V_{CC} = 4.5 V	3.15			3.15			
		V_{CC} = 6 V	4.2			4.2			
V_{IL}	Low-level input voltage	V_{CC} = 2 V	0		0.3	0		0.3	V
		V_{CC} = 4.5 V	0		0.9	0		0.9	
		V_{CC} = 6 V	0		1.2	0		1.2	
V_I	Input voltage		0		V_{CC}	0		V_{CC}	V
V_O	Output voltage		0		V_{CC}	0		V_{CC}	V
t_t	Input transition (rise and fall) times	V_{CC} = 2 V	0		1000	0		1000	ns
		V_{CC} = 4.5 V	0		500	0		500	
		V_{CC} = 6 V	0		400	0		400	
T_A	Operating free-air temperature		-55		125	-40		85	°C

electrical characteristics over recommended operating free-air temperature range (unless otherwise noted)

PARAMETER	TEST CONDITIONS	V_{CC}	T_A = 25 °C			SN54HC00		SN74HC00		UNIT
			MIN	TYP	MAX	MIN	MAX	MIN	MAX	
V_{OH}	$V_I = V_{IH}$ or V_{IL}, $I_{OH} = -20$ μA	2 V	1.9	1.998		1.9		1.9		V
		4.5 V	4.4	4.499		4.4		4.4		
		6 V	5.9	5.999		5.9		5.9		
	$V_I = V_{IH}$ or V_{IL}, $I_{OH} = -4$ mA	4.5 V	3.98	4.30		3.7		3.84		
	$V_I = V_{IH}$ or V_{IL}, $I_{OH} = -5.2$ mA	6 V	5.48	5.80		5.2		5.34		
V_{OL}	$V_I = V_{IH}$ or V_{IL}, $I_{OL} = 20$ μA	2 V		0.002	0.1		0.1		0.1	V
		4.5 V		0.001	0.1		0.1		0.1	
		6 V		0.001	0.1		0.1		0.1	
	$V_I = V_{IH}$ or V_{IL}, $I_{OL} = 4$ mA	4.5 V		0.17	0.26		0.4		0.33	
	$V_I = V_{IH}$ or V_{IL}, $I_{OL} = 5.2$ mA	6 V		0.15	0.26		0.4		0.33	
I_I	$V_I = V_{CC}$ or 0	6 V		± 0.1	± 100		± 1000		± 1000	nA
I_{CC}	$V_I = V_{CC}$ or 0, $I_O = 0$	6 V			2		40		20	μA
C_i		2 to 6 V		3	10		10		10	pF

TEXAS
INSTRUMENTS

POST OFFICE BOX 655012 • DALLAS, TEXAS 75265

switching characteristics over recommended operating free-air temperature range (unless otherwise noted), C_L = 50 pF (see Note 1)

PARAMETER	FROM (INPUT)	TO (OUTPUT)	V_{CC}	T_A = 25°C			SN54HC00		SN74HC00		UNIT
				MIN	TYP	MAX	MIN	MAX	MIN	MAX	
t_{pd}	A or B	Y	2 V		45	90		135		115	
			4.5 V		9	18		27		23	ns
			6 V		8	15		23		20	
t_t		Y	2 V		38	75		110		95	
			4.5 V		8	15		22		19	ns
			6 V		6	13		19		16	

C_{pd}	Power dissipation capacitance per gate	No load, T_A = 25°C	20 pF typ

NOTE 1: Load circuit and voltage waveforms are shown in Section 1.

2

HCMOS Devices

- Package Options Include Plastic "Small Outline" Packages, Ceramic Chip Carriers and Flat Packages, and Plastic and Ceramic DIPs

- Dependable Texas Instruments Quality and Reliability

description

These devices contain four independent 2-input-NOR gates.

The SN5402, SN54LS02, and SN54S02 are characterized for operation over the full military temperature range of −55°C to 125°C. The SN7402, SN74LS02, and SN74S02 are characterized for operation from 0°C to 70°C.

FUNCTION TABLE (each gate)

INPUTS		OUTPUT
A	**B**	**Y**
H	X	L
X	H	L
L	L	H

logic symbol†

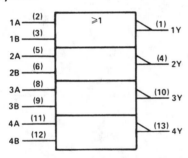

†This symbol is in accordance with ANSI/IEEE Std. 91-1984 and IEC Publication 617-12.
Pin numbers shown are for D, J, and N packages.

logic diagram (positive logic)

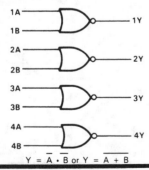

SN5402 . . . J PACKAGE
SN54LS02, SN54S02 . . . J OR W PACKAGE
SN7402 . . . N PACKAGE
SN74LS02, SN74S02 . . . D OR N PACKAGE
(TOP VIEW)

```
1Y  [1    14] VCC
1A  [2    13] 4Y
1B  [3    12] 4B
2Y  [4    11] 4A
2A  [5    10] 3Y
2B  [6     9] 3B
GND [7     8] 3A
```

SN5402 . . . W PACKAGE
(TOP VIEW)

```
1A  [1    14] 4Y
1B  [2    13] 4B
1Y  [3    12] 4A
VCC [4    11] GND
2Y  [5    10] 3B
2A  [6     9] 3A
2B  [7     8] 3Y
```

SN54LS02, SN54S02 . . . FK PACKAGE
(TOP VIEW)

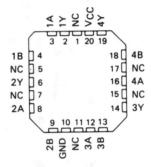

NC - No internal connection

TEXAS INSTRUMENTS
POST OFFICE BOX 655012 • DALLAS, TEXAS 75265

2

TTL Devices

SN5402, SN54LS02, SN54S02,
SN7402, SN74LS02, SN74S02
QUADRUPLE 2-INPUT POSITIVE-NOR GATES

schematics (each gate)

'02

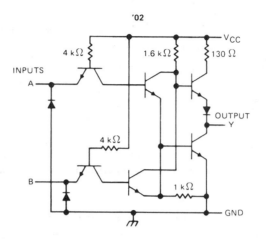

'LS02

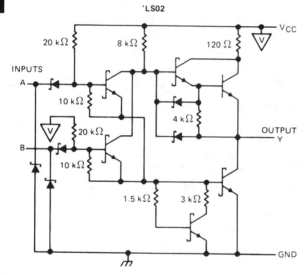

'S02

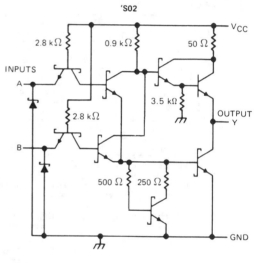

Resistor values shown are nominal.

absolute maximum ratings over operating free-air temperature range (unless otherwise noted)

Supply voltage, V_{CC} (see Note 1) . . . . 7 V
Input voltage: '02, 'S02 . . . . 5.5 V
               'LS02 . . . . 7 V
Off-state output voltage . . . . 7 V
Operating free-air temperature range: SN54' . . . . −55°C to 125°C
                                SN74' . . . . 0°C to 70°C
Storage temperature range . . . . −65°C to 150°C

NOTE 1. Voltage values are with respect to network ground terminal.

TEXAS
INSTRUMENTS
POST OFFICE BOX 655012 • DALLAS, TEXAS 75265

SN54LS02, SN74LS02
QUADRUPLE 2-INPUT POSITIVE-NOR GATES

recommended operating conditions

		SN54LS02			SN74LS02			UNIT
		MIN	NOM	MAX	MIN	NOM	MAX	
V_{CC}	Supply voltage	4.5	5	5.5	4.75	5	5.25	V
V_{IH}	High-level input voltage	2			2			V
V_{IL}	Low-level input voltage			0.7			0.8	V
I_{OH}	High-level output current			−0.4			−0.4	mA
I_{OL}	Low-level output current			4			8	mA
T_A	Operating free-air temperature	−55		125	0		70	°C

electrical characteristics over recommended operating free-air temperature range (unless otherwise noted)

PARAMETER	TEST CONDITIONS †			SN54LS02			SN74LS02			UNIT
			MIN	TYP‡	MAX	MIN	TYP‡	MAX		
V_{IK}	V_{CC} = MIN,	I_I = −18 mA			−1.5			−1.5		V
V_{OH}	V_{CC} = MIN,	V_{IL} = MAX, I_{OH} = −0.4 mA	2.5	3.4		2.7	3.4			V
V_{OL}	V_{CC} = MIN,	V_{IH} = 2 V, I_{OL} = 4 mA		0.25	0.4		0.25	0.4		V
	V_{CC} = MIN,	V_{IH} = 2 V, I_{OL} = 8 mA					0.35	0.5		
I_I	V_{CC} = MAX,	V_I = 7 V			0.1			0.1		mA
I_{IH}	V_{CC} = MAX,	V_I = 2.7 V			20			20		μA
I_{IL}	V_{CC} = MAX,	V_I = 0.4 V			−0.4			−0.4		mA
I_{OS}§	V_{CC} = MAX		−20		−100	−20		−100		mA
I_{CCH}	V_{CC} = MAX,	V_I = 0 V		1.6	3.2		1.6	3.2		mA
I_{CCL}	V_{CC} = MAX,	See Note 2		2.8	5.4		2.8	5.4		mA

† For conditions shown as MIN or MAX, use the appropriate value specified under recommended operating conditions.
‡ All typical values are at V_{CC} = 5 V, T_A = 25°C.
§ Not more than one output should be shorted at a time, and the duration of the short-circuit should not exceed one second.
NOTE 2: One input at 4.5 V, all others at GND.

switching characteristics, V_{CC} = 5 V, T_A = 25°C (see note 3)

PARAMETER	FROM (INPUT)	TO (OUTPUT)	TEST CONDITIONS		MIN	TYP	MAX	UNIT
t_{PLH}	A or B	Y	R_L = 2 kΩ,	C_L = 15 pF		10	15	ns
t_{PHL}						10	15	ns

NOTE 3: Load circuits and voltage waveforms are shown in Section 1.

TEXAS
INSTRUMENTS
POST OFFICE BOX 655012 • DALLAS, TEXAS 75265

- **Package Options Include Plastic "Small Outline" Packages, Ceramic Chip Carriers and Flat Packages, and Plastic and Ceramic DIPs**

- **Dependable Texas Instruments Quality and Reliability**

description

These devices contain six independent inverters.

The SN5404, SN54LS04, and SN54S04 are characterized for operation over the full military temperature range of −55°C to 125°C. The SN7404, SN74LS04, and SN74S04 are characterized for operation from 0°C to 70°C.

FUNCTION TABLE (each inverter)

INPUTS A	OUTPUT Y
H	L
L	H

logic symbol†

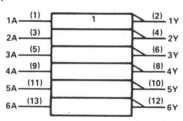

†This symbol is in accordance with ANSI/IEEE Std. 91-1984 and IEC Publication 617-12.
Pin numbers shown are for D, J, and N packages.

logic diagram (positive logic)

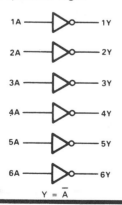

$$Y = \overline{A}$$

SN5404 . . . J PACKAGE
SN54LS04, SN54S04 . . . J OR W PACKAGE
SN7404 . . . N PACKAGE
SN74LS04, SN74S04 . . . D OR N PACKAGE
(TOP VIEW)

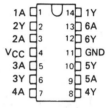

1A 1	14 V_CC
1Y 2	13 6A
2A 3	12 6Y
2Y 4	11 5A
3A 5	10 5Y
3Y 6	9 4A
GND 7	8 4Y

SN5404 . . . W PACKAGE
(TOP VIEW)

1A 1	14 1Y
2Y 2	13 6A
2A 3	12 6Y
V_CC 4	11 GND
3A 5	10 5Y
3Y 6	9 5A
4A 7	8 4Y

SN54LS04, SN54S04 . . . FK PACKAGE
(TOP VIEW)

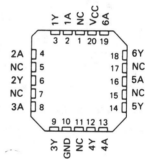

NC - No internal connection

TEXAS
INSTRUMENTS

POST OFFICE BOX 655012 • DALLAS, TEXAS 75265

SN5404, SN54LS04, SN54S04,
SN7404, SN74LS04, SN74S04
HEX INVERTERS

schematics (each gate)

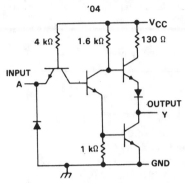

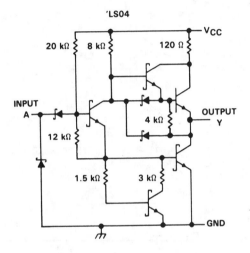

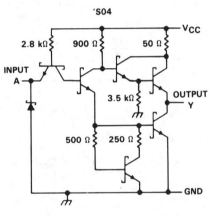

Resistor values shown are nominal.

absolute maximum ratings over operating free-air temperature range (unless otherwise noted)

Supply voltage, V_{CC} (see Note 1) . . . . 7 V
Input voltage: '04, 'S04 . . . . 5.5 V
               'LS04 . . . . 7 V
Operating free-air temperature range: SN54' . . . . −55°C to 125°C
                             SN74' . . . . 0°C to 70°C
Storage temperature range . . . . −65°C to 150°C

NOTE 1: Voltage values are with respect to network ground terminal.

TEXAS
INSTRUMENTS
POST OFFICE BOX 655012 • DALLAS, TEXAS 75265

SN54LS04, SN74LS04
HEX INVERTERS

recommended operating conditions

		SN54LS04 MIN	NOM	MAX	SN74LS04 MIN	NOM	MAX	UNIT
V_{CC}	Supply voltage	4.5	5	5.5	4.75	5	5.25	V
V_{IH}	High-level input voltage	2			2			V
V_{IL}	Low-level input voltage			0.7			0.8	V
I_{OH}	High-level output current			− 0.4			− 0.4	mA
I_{OL}	Low-level output current			4			8	mA
T_A	Operating free-air temperature	− 55		125	0		70	°C

electrical characteristics over recommended operating free-air temperature range (unless otherwise noted)

PARAMETER	TEST CONDITIONS †	SN54LS04 MIN	TYP‡	MAX	SN74LS04 MIN	TYP‡	MAX	UNIT
V_{IK}	V_{CC} = MIN, I_I = − 18 mA			− 1.5			− 1.5	V
V_{OH}	V_{CC} = MIN, V_{IL} = MAX, I_{OH} = − 0.4 mA	2.5	3.4		2.7	3.4		V
V_{OL}	V_{CC} = MIN, V_{IH} = 2 V, I_{OL} = 4 mA		0.25	0.4			0.4	V
	V_{CC} = MIN, V_{IH} = 2 V, I_{OL} = 8 mA					0.25	0.5	
I_I	V_{CC} = MAX, V_I = 7 V			0.1			0.1	mA
I_{IH}	V_{CC} = MAX, V_I = 2.7 V			20			20	μA
I_{IL}	V_{CC} = MAX, V_I = 0.4 V			− 0.4			− 0.4	mA
I_{OS} §	V_{CC} = MAX	− 20		− 100	− 20		− 100	mA
I_{CCH}	V_{CC} = MAX, V_I = 0 V		1.2	2.4		1.2	2.4	mA
I_{CCL}	V_{CC} = MAX, V_I = 4.5 V		3.6	6.6		3.6	6.6	mA

† For conditions shown as MIN or MAX, use the appropriate value specified under recommended operating conditions.
‡ All typical values are at V_{CC} = 5 V, T_A = 25°C.
§ Not more than one output should be shorted at a time, and the duration of the short-circuit should not exceed one second.

switching characteristics, V_{CC} = 5 V, T_A = 25°C (see note 2)

PARAMETER	FROM (INPUT)	TO (OUTPUT)	TEST CONDITIONS	MIN	TYP	MAX	UNIT
t_{PLH}	A	Y	R_L = 2 kΩ, C_L = 15 pF		9	15	ns
t_{PHL}					10	15	ns

NOTE 2: Load circuits and voltage waveforms are shown in Section 1.

TEXAS
INSTRUMENTS
POST OFFICE BOX 655012 • DALLAS, TEXAS 75265

- Package Options Include Plastic "Small Outline" Packages, Ceramic Chip Carriers and Flat Packages, and Plastic and Ceramic DIPs

- Dependable Texas Instruments Quality and Reliability

description

These devices contain four independent 2-input AND gates.

The SN5408, SN54LS08, and SN54S08 are characterized for operation over the full military temperature range of −55°C to 125°C. The SN7408, SN74LS08 and SN74S08 are characterized for operation from 0° to 70°C.

FUNCTION TABLE (each gate)

INPUTS		OUTPUT
A	B	Y
H	H	H
L	X	L
X	L	L

logic symbol[†]

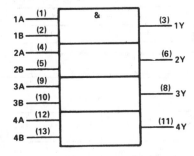

[†] This symbol is in accordance with ANSI/IEEE Std 91-1984 and IEC Publication 617-12.
Pin numbers shown are for D, J, N, and W packages.

SN5408, SN54LS08, SN54S08 . . . J OR W PACKAGE
SN7408 . . . J OR N PACKAGE
SN74LS08, SN74S08 . . . D, J OR N PACKAGE
(TOP VIEW)

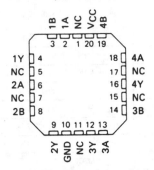

SN54LS08, SN54S08 . . . FK PACKAGE
(TOP VIEW)

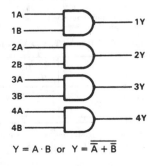

NC—No internal connection

logic diagram (positive logic)

1A, 1B → 1Y
2A, 2B → 2Y
3A, 3B → 3Y
4A, 4B → 4Y

$$Y = A \cdot B \quad \text{or} \quad Y = \overline{\overline{A} + \overline{B}}$$

TEXAS INSTRUMENTS
POST OFFICE BOX 655012 • DALLAS, TEXAS 75265

2

TTL Devices

schematics (each gate)

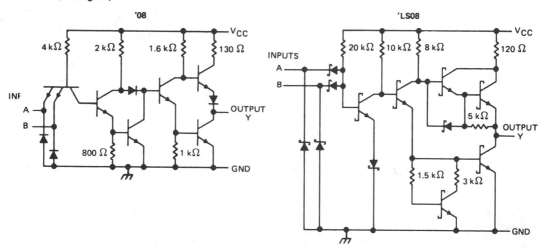

'08 · 'LS08

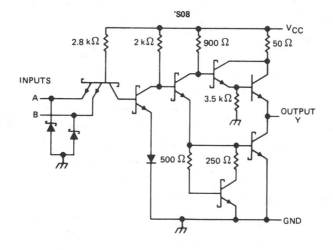

'S08

Resistor values are nominal.

absolute maximum ratings over operating free-air temperature range (unless otherwise noted)

Supply voltage, V_{CC} (see Note 1) . . . . 7 V
Input voltage: '08, 'S08 . . . . 5.5 V
'LS08 . . . . 7 V
Operating free-air temperature range: SN54' . . . . −55°C to 125°C
SN74' . . . . 0°C to 70°C
Storage temperature range . . . . −65°C to 150°C

NOTE 1: Voltage values are with respect to network ground terminal.

**TEXAS
INSTRUMENTS**
POST OFFICE BOX 655012 • DALLAS, TEXAS 75265

recommended operating conditions

		SN54LS08			SN74LS08			UNIT
		MIN	NOM	MAX	MIN	NOM	MAX	
V_{CC}	Supply voltage	4.5	5	5.5	4.75	5	5.25	V
V_{IH}	High-level input voltage	2			2			V
V_{IL}	Low-level input voltage			0.7			0.8	V
I_{OH}	High-level output current			−0.4			−0.4	mA
I_{OL}	Low-level output current			4			8	mA
T_A	Operating free-air temperature	−55		125	0		70	°C

electrical characteristics over recommended operating free-air temperature range (unless otherwise noted)

PARAMETER	TEST CONDITIONS †		SN54LS08			SN74LS08			UNIT
			MIN	TYP‡	MAX	MIN	TYP‡	MAX	
V_{IK}	V_{CC} = MIN,	I_I = −18 mA			−1.5			−1.5	V
V_{OH}	V_{CC} = MIN, V_{IH} = 2 V,	I_{OH} = −0.4 mA	2.5	3.4		2.7	3.4		V
V_{OL}	V_{CC} = MIN, V_{IL} = MAX,	I_{OL} = 4 mA		0.25	0.4		0.25	0.4	V
	V_{CC} = MIN, V_{IL} = MAX,	I_{OL} = 8 mA					0.35	0.5	
I_I	V_{CC} = MAX,	V_I = 7 V			0.1			0.1	mA
I_{IH}	V_{CC} = MAX,	V_I = 2.7 V			20			20	µA
I_{IL}	V_{CC} = MAX,	V_I = 0.4 V			−0.4			−0.4	mA
I_{OS}§	V_{CC} = MAX		−20		−100	−20		−100	mA
I_{CCH}	V_{CC} = MAX,	V_I = 4.5 V		2.4	4.8		2.4	4.8	mA
I_{CCL}	V_{CC} = MAX,	V_I = 0 V		4.4	8.8		4.4	8.8	mA

† For conditions shown as MIN or MAX, use the appropriate value specified under recommended operating conditions.
‡ All typical values are at V_{CC} = 5 V, T_A = 25°C
§ Not more than one output should be shorted at a time, and the duration of the short-circuit should not exceed one second.

switching characteristics, V_{CC} = 5 V, T_A = 25°C (see note 2)

PARAMETER	FROM (INPUT)	TO (OUTPUT)	TEST CONDITIONS		MIN	TYP	MAX	UNIT
t_{PLH}	A or B	Y	R_L = 2 kΩ,	C_L = 15 pF		8	15	ns
t_{PHL}						10	20	ns

NOTE 2: Load circuits and voltage waveforms are shown in Section 1.

TEXAS
INSTRUMENTS

POST OFFICE BOX 655012 • DALLAS, TEXAS 75265

- Package Options Include Plastic "Small Outline" Packages, Ceramic Chip Carriers and Flat Packages, and Plastic and Ceramic DIPs

- Dependable Texas Instruments Quality and Reliability

description

These devices contain three independent 3-input NAND gates.

The SN5410, SN54LS10, and SN54S10 are characterized for operation over the full military temperature range of −55°C to 125°C. The SN7410, SN74LS10, and SN74S10 are characterized for operation from 0°C to 70°C.

FUNCTION TABLE (each gate)

INPUTS			OUTPUT
A	B	C	Y
H	H	H	L
L	X	X	H
X	L	X	H
X	X	L	H

logic symbol†

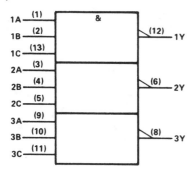

†This symbol is in accordance with ANSI/IEEE Std. 91-1984 and IEC Publication 617-12.

Pin numbers shown are for D, J, and N packages.

positive logic

$$Y = \overline{A \cdot B \cdot C} \text{ or } Y = \overline{A} + \overline{B} + \overline{C}$$

SN5410 . . . J PACKAGE
SN54LS10, SN54S10 . . . J OR W PACKAGE
SN7410 . . . N PACKAGE
SN74LS10, SN74S10 . . . D OR N PACKAGE
(TOP VIEW)

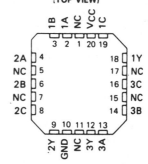

```
1A  [1    14]  VCC
1B  [2    13]  1C
2A  [3    12]  1Y
2B  [4    11]  3C
2C  [5    10]  3B
2Y  [6     9]  3A
GND [7     8]  3Y
```

SN5410 . . . W PACKAGE
(TOP VIEW)

```
1A   [1    14]  1C
1B   [2    13]  3Y
1Y   [3    12]  3C
VCC  [4    11]  GND
2Y   [5    10]  3B
2A   [6     9]  3A
2B   [7     8]  2C
```

SN54LS10, SN54S10 . . . FK PACKAGE
(TOP VIEW)

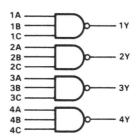

NC - No internal connection

logic diagram (positive logic)

```
1A ─┐
1B ─┤ ◯─ 1Y
1C ─┘
2A ─┐
2B ─┤ ◯─ 2Y
2C ─┘
3A ─┐
3B ─┤ ◯─ 3Y
3C ─┘
4A ─┐
4B ─┤ ◯─ 4Y
4C ─┘
```

2

TTL Devices

TEXAS
INSTRUMENTS
POST OFFICE BOX 655012 • DALLAS, TEXAS 75265

- **Operation from Very Slow Edges**
- **Improved Line-Receiving Characteristics**
- **High Noise Immunity**

description

Each circuit functions as an inverter, but because of the Schmitt action, it has different input threshold levels for positive (V_{T+}) and for negative going (V_{T-}) signals.

These circuits are temperature-compensated and can be triggered from the slowest of input ramps and still give clean, jitter-free output signals.

The SN5414 and SN54LS14 are characterized for operation over the full military temperature range of –55°C to 125°C. The SN7414 and the SN74LS14 are characterized for operation from 0°C to 70°C.

logic symbol[†]

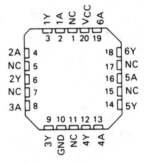

SN5414, SN54LS14 . . . J OR W PACKAGE
SN7414 . . . N PACKAGE
SN74LS14 . . . D OR N PACKAGE
(TOP VIEW)

1A	1	14	V_{CC}
1Y	2	13	6A
2A	3	12	6Y
2Y	4	11	5A
3A	5	10	5Y
3Y	6	9	4A
GND	7	8	4Y

SN54LS14 . . . FK PACKAGE
(TOP VIEW)

NC – No internal connection

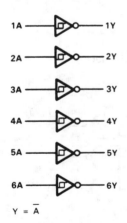

[†] This symbol is in accordance with ANSI/IEEE Std 91-1984 and IEC Publication 617-12.
Pin numbers shown are for D, J, N, and W packages.

logic diagram (positive logic)

$Y = \overline{A}$

TEXAS
INSTRUMENTS

POST OFFICE BOX 655012 • DALLAS, TEXAS 75265

TTL Devices

2

2-77

SN5414, SN54LS14, SN7414, SN74LS14
HEX SCHMITT-TRIGGER INVERTERS

schematics

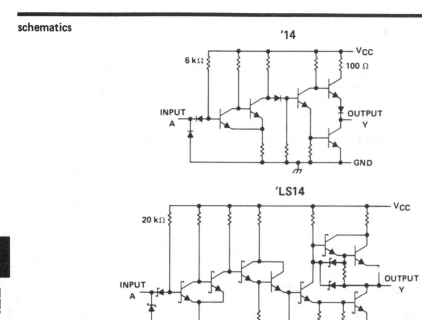

'14

'LS14

Resistor values shown are nominal.

absolute maximum ratings over operating free-air temperature range (unless otherwise noted)

Supply voltage, V_{CC} (see Note 1)	7 V
Input voltage: '14	5.5 V
'LS14	7 V
Operating free-air temperature: SN54'	-55°C to 125°C
SN74'	0°C to 70°C
Storage temperature range	-65°C to 150°C

NOTE 1: Voltage values are with respect to network ground terminal.

2 TTL Devices

TEXAS
INSTRUMENTS

POST OFFICE BOX 655012 • DALLAS, TEXAS 75265

recommended operating conditions

		SN54LS14			SN74LS14			UNIT
		MIN	NOM	MAX	MIN	NOM	MAX	
V_{CC}	Supply voltage	4.5	5	5.5	4.75	5	5.25	V
I_{OH}	High-level output current			−0.4			−0.4	mA
I_{OL}	Low-level output current			4			8	mA
T_A	Operating free-air temperature	−55		125	0		70	°C

electrical characteristics over recommended operating free-air temperature range (unless otherwise noted)

PARAMETER	TEST CONDITIONS†			SN54LS14			SN74LS14			UNIT
				MIN	TYP‡	MAX	MIN	TYP‡	MAX	
V_{T+}	V_{CC} = 5 V			1.4	1.6	1.9	1.4	1.6	1.9	V
V_{T-}	V_{CC} = 5 V			0.5	0.8	1	0.5	0.8	1	V
Hysteresis ($V_{T+} - V_{T-}$)	V_{CC} = 5 V			0.4	0.8		0.4	0.8		V
V_{IK}	V_{CC} = MIN,	I_I = −18 mA				−1.5			−1.5	V
V_{OH}	V_{CC} = MIN,	V_I = 0.5 V,	I_{OH} = −0.4 mA	2.5	3.4		2.7	3.4		V
V_{OL}	V_{CC} = MIN,	V_I = 1.9 V	I_{OL} = 4 mA		0.25	0.4		0.25	0.4	V
			I_{OL} = 8 mA					0.35	0.5	
I_{T+}	V_{CC} = 5 V,	V_I = V_{T+}			−0.14			−0.14		mA
I_{T-}	V_{CC} = 5 V,	V_I = V_{T-}			−0.18			−0.18		mA
I_I	V_{CC} = MAX,	V_I = 7 V				0.1			0.1	mA
I_{IH}	V_{CC} = MAX,	V_{IH} = 2.7 V				20			20	µA
I_{IL}	V_{CC} = MAX,	V_{IL} = 0.4 V				−0.4			−0.4	mA
I_{OS}§	V_{CC} = MAX			−20		−100	−20		−100	mA
I_{CCH}	V_{CC} = MAX				8.6	16		8.6	16	mA
I_{CCL}	V_{CC} = MAX				12	21		12	21	mA

† For conditions shown as MIN or MAX, use the appropriate value specified under recommended operating conditions.
‡ All typical values are at V_{CC} = 5 V, T_A = 25°C.
§ Not more than one output should be shorted at a time, and duration of the short-circuit should not exceed one second.

switching characteristics, V_{CC} = 5 V, T_A = 25°C

PARAMETER	FROM (INPUT)	TO (OUTPUT)	TEST CONDITIONS		MIN	TYP	MAX	UNIT
t_{PLH}	A	Y	R_L = 2 kΩ,	C_L = 15 pF		15	22	ns
t_{PHL}						15	22	ns

TEXAS
INSTRUMENTS
POST OFFICE BOX 655012 • DALLAS, TEXAS 75265

TYPICAL CHARACTERISTICS OF 'LS14 CIRCUITS

POSITIVE-GOING THRESHOLD VOLTAGE
vs
FREE-AIR TEMPERATURE

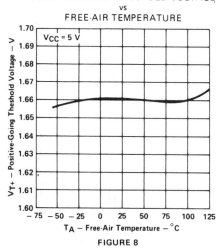

FIGURE 8

NEGATIVE-GOING THRESHOLD VOLTAGE
vs
FREE-AIR TEMPERATURE

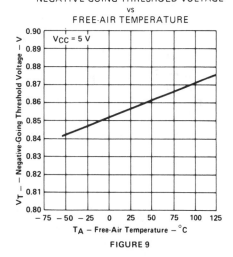

FIGURE 9

HYSTERESIS
vs
FREE-AIR TEMPERATURE

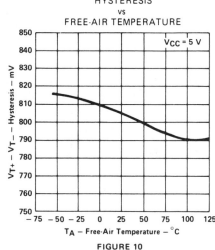

FIGURE 10

DISTRIBUTION OF UNITS
FOR HYSTERESIS

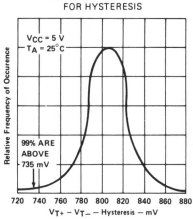

FIGURE 11

Data for temperatures below 0°C and above 70°C and supply voltages below 4.75 V and above 5.25 V are applicable for SN54LS14 only.

TTL Devices

2

SN54LS14, SN74LS14
HEX SCHMITT-TRIGGER INVERTERS

TYPICAL CHARACTERISTICS OF 'LS14 CIRCUITS

THRESHOLD VOLTAGES AND HYSTERESIS
vs
SUPPLY VOLTAGE

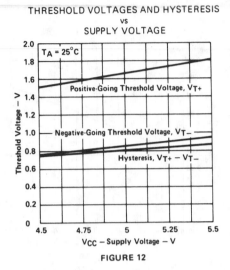

FIGURE 12

OUTPUT VOLTAGE
vs
INPUT VOLTAGE

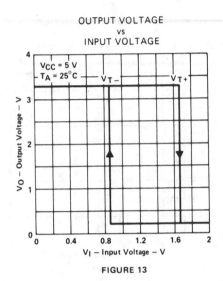

FIGURE 13

Data for temperatures below 0°C and above 70°C and supply voltages below 4.75 V and above 5.25 V are applicable for SN54LS14 only.

TEXAS
INSTRUMENTS
POST OFFICE BOX 655012 • DALLAS, TEXAS 75265

TYPICAL APPLICATION DATA

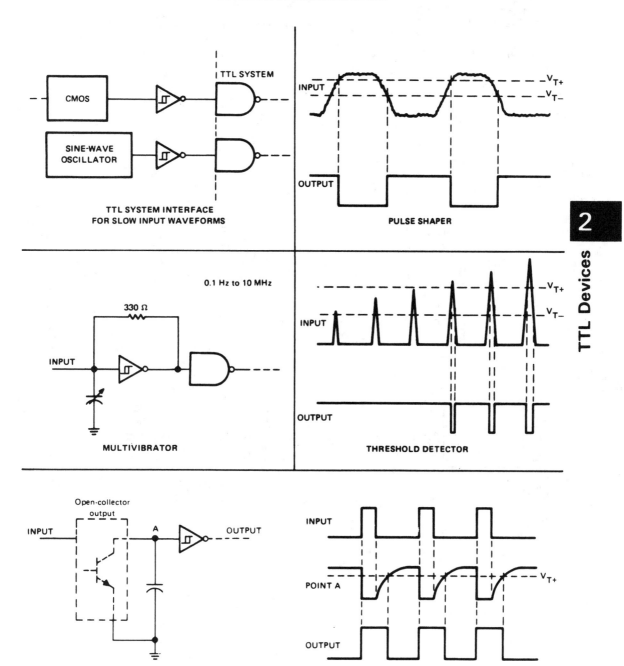

TTL SYSTEM INTERFACE
FOR SLOW INPUT WAVEFORMS

PULSE SHAPER

MULTIVIBRATOR

THRESHOLD DETECTOR

PULSE STRETCHER

2

TTL Devices

TEXAS
INSTRUMENTS
POST OFFICE BOX 655012 • DALLAS, TEXAS 75265

- Package Options Include Plastic "Small Outline" Packages, Ceramic Chip Carriers and Flat Packages, and Plastic and Ceramic DIPs

- Dependable Texas Instruments Quality and Reliability

description

These devices contain two independent 4-input NAND gates.

The SN5420, SN54LS20, and SN54S20 are characterized for operation over the full military range of −55°C to 125°C. The SN7420, SN74LS20, and SN74S20 are characterized for opertion from 0°C to 70°C.

FUNCTION TABLE (each gate)

INPUTS				OUTPUT
A	B	C	D	Y
H	H	H	H	L
L	X	X	X	H
X	L	X	X	H
X	X	L	X	H
X	X	X	L	H

logic symbol[†]

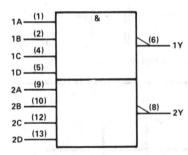

1A (1)
1B (2)
1C (4)
1D (5)
&
(6) 1Y

2A (9)
2B (10)
2C (12)
2D (13)
(8) 2Y

[†]This symbol is in accordance with ANSI/IEEE Std. 91-1984 and IEC Publication 617-12.

Pin numbers shown are for D, J, N, and W packages.

SN5420 . . . J PACKAGE
SN54LS20, SN54S20 . . . J OR W PACKAGE
SN7420 . . . N PACKAGE
SN74LS20, SN74S20 . . . D OR N PACKAGE
(TOP VIEW)

```
1A   1   14  Vcc
1B   2   13  2D
NC   3   12  2C
1C   4   11  NC
1D   5   10  2B
1Y   6    9  2A
GND  7    8  2Y
```

SN5420 . . . W PACKAGE
(TOP VIEW)

```
1A   1   14  1D
1Y   2   13  1C
NC   3   12  1B
Vcc  4   11  GND
NC   5   10  2Y
2A   6    9  2D
2B   7    8  2C
```

SN54LS20, SN54S20 . . . FK PACKAGE
(TOP VIEW)

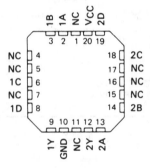

```
            1B 1A NC Vcc 2D
             3  2  1 20 19
NC    4               18  2C
NC    5               17  NC
1C    6               16  NC
NC    7               15  NC
1D    8               14  2B
             9 10 11 12 13
            1Y GND NC 2Y 2A
```

NC - No internal connection

logic diagram

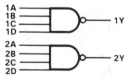

1A
1B
1C
1D
1Y

2A
2B
2C
2D
2Y

positive logic $Y = \overline{A \cdot B \cdot C \cdot D}$ or $Y = \overline{A} + \overline{B} + \overline{C} + \overline{D}$

2

TTL Devices

TEXAS
INSTRUMENTS

POST OFFICE BOX 655012 • DALLAS, TEXAS 75265

DECEMBER 1983 - REVISED MARCH 1988

- **Package Options Include Plastic "Small Outline" Packages, Ceramic Chip Carriers and Flat Packages, and Plastic and Ceramic DIPs**

- **Dependable Texas Instruments Quality and Reliability**

description

These devices contain three independent 3-input NOR gates.

The SN5427 and SN54LS27 are characterized for operation over the full military temperature range of −55°C to 125°C. The SN7427 and SN74LS27 are characterized for operation from 0°C to 70°C.

FUNCTION TABLE (each gate)

INPUTS			OUTPUT
A	B	C	Y
H	X	X	L
X	H	X	L
X	X	H	L
L	L	L	H

logic symbol†

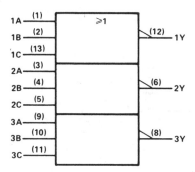

† This symbol is in accordance with ANSI/IEEE Std 91-1984 and IEC Publication 617-12.

Pin numbers shown are for D, J, N, and W packages

SN5427, SN54LS27 . . . J OR W PACKAGE
SN7427 . . . N PACKAGE
SN74LS27 . . . D OR N PACKAGE
(TOP VIEW)

```
        ┌───┬─┬───┐
  1A  [1│   U   │14]  VCC
  1B  [2│       │13]  1C
  2A  [3│       │12]  1Y
  2B  [4│       │11]  3C
  2C  [5│       │10]  3B
  2Y  [6│       │ 9]  3A
 GND  [7│       │ 8]  3Y
        └───────┘
```

SN54LS27 . . . FK PACKAGE
(TOP VIEW)

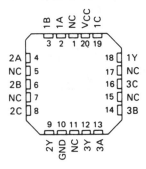

NC - No internal connection

logic diagram

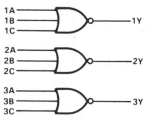

positive logic

$$Y = \overline{A + B + C} \text{ or } Y = \overline{A} \cdot \overline{B} \cdot \overline{C}$$

TEXAS
INSTRUMENTS

POST OFFICE BOX 225012 • DALLAS, TEXAS 75265

2

TTL Devices

- Package Options Include Plastic "Small Outline" Packages, Ceramic Chip Carriers and Flat Packages, and Plastic and Ceramic DIPs

- Dependable Texas Instruments Quality and Reliability

description

These devices contain four independent 2-input OR gates.

The SN5432, SN54LS32 and SN54S32 are characterized for operation over the full military range of −55°C to 125°C. The SN7432, SN74LS32 and SN74S32 are characterized for operation from 0°C to 70°C.

FUNCTION TABLE (each gate)

INPUTS		OUTPUT
A	B	Y
H	X	H
X	H	H
L	L	L

logic symbol†

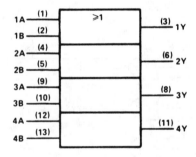

† This symbol is in accordance with ANSI IEEE Std 91 1984 and IEC Publication 617 12.

Pin numbers shown are for D. J. N. or W packages

SN5432, SN54LS32, SN54S32 . . . J OR W PACKAGE
SN7432 . . . N PACKAGE
SN74LS32, SN74S32 . . . D OR N PACKAGE
(TOP VIEW)

```
      ┌───┬─┬───┐
 1A  │1  U  14│  VCC
 1B  │2      13│  4B
 1Y  │3      12│  4A
 2A  │4      11│  4Y
 2B  │5      10│  3B
 2Y  │6       9│  3A
 GND │7       8│  3Y
      └─────────┘
```

SN54LS32, SN54S32 . . . FK PACKAGE
(TOP VIEW)

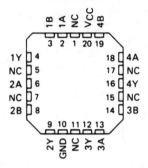

NC - No internal connection

logic diagram

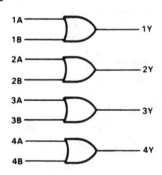

positive logic

$$Y = A + B \text{ or } Y = \overline{\overline{A} \cdot \overline{B}}$$

2

TTL Devices

TEXAS
INSTRUMENTS

POST OFFICE BOX 655012 • DALLAS, TEXAS 75265

'46A, '47A, 'LS47 feature	'48, 'LS48 feature	'LS49 feature
• **Open-Collector Outputs Drive Indicators Directly** • **Lamp-Test Provision** • **Leading/Trailing Zero Suppression**	• **Internal Pull-Ups Eliminate Need for External Resistors** • **Lamp-Test Provision** • **Leading/Trailing Zero Suppression**	• **Open-Collector Outputs** • **Blanking Input**

SN5446A, SN5447A, SN54LS47, SN5448,
SN54LS48 . . . J PACKAGE
SN7446A, SN7447A,
SN7448 . . . N PACKAGE
SN74LS47, SN74LS48 . . . D OR N PACKAGE
(TOP VIEW)

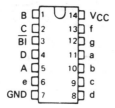

SN54LS47, SN54LS48 . . . FK PACKAGE
(TOP VIEW)

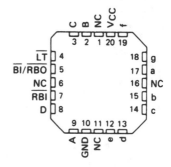

SN54LS49 . . . J OR W PACKAGE
SN74LS49 . . . D OR N PACKAGE
(TOP VIEW)

SN54LS49 . . . FK PACKAGE
(TOP VIEW)

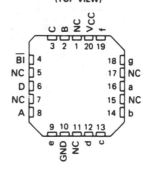

NC – No internal connection

2

TTL Devices

TEXAS
INSTRUMENTS

POST OFFICE BOX 655012 • DALLAS, TEXAS 75265

SN5446A, '47A, '48, SN54LS47, 'LS48, 'LS49, SN7446A, '47A, '48, SN74LS47, 'LS48, 'LS49 BCD-TO-SEVEN-SEGMENT DECODERS/DRIVERS

- All Circuit Types Feature Lamp Intensity Modulation Capability

TYPE	DRIVER OUTPUTS				TYPICAL POWER DISSIPATION	PACKAGES
	ACTIVE LEVEL	OUTPUT CONFIGURATION	SINK CURRENT	MAX VOLTAGE		
SN5446A	low	open-collector	40 mA	30 V	320 mW	J, W
SN5447A	low	open-collector	40 mA	15 V	320 mW	J, W
SN5448	high	2-kΩ pull-up	6.4 mA	5.5 V	265 mW	J, W
SN54LS47	low	open-collector	12 mA	15 V	35 mW	J, W
SN54LS48	high	2-kΩ pull-up	2 mA	5.5 V	125 mW	J, W
SN54LS49	high	open-collector	4 mA	5.5 V	40 mW	J, W
SN7446A	low	open-collector	40 mA	30 V	320 mW	J, N
SN7447A	low	open-collector	40 mA	15 V	320 mW	J, N
SN7448	high	2-kΩ pull-up	6.4 mA	5.5 V	265 mW	J, N
SN74LS47	low	open-collector	24 mA	15 V	35 mW	J, N
SN74LS48	high	2-kΩ pull-up	6 mA	5.5 V	125 mW	J, N
SN74LS49	high	open-collector	8 mA	5.5 V	40 mW	J, N

logic symbols[†]

'46A, '47A, 'LS47

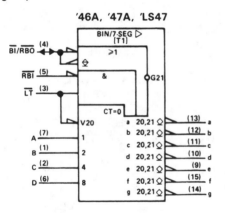

'48, 'LS48

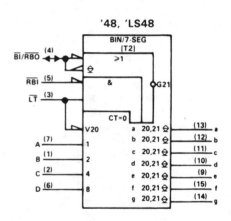

'LS49

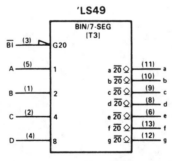

[†]These symbols are in accordance with ANSI/IEEE Std 91-1984 and IEC Publication 617-12.
Pin numbers shown are for D, J, N, and W packages.

TEXAS INSTRUMENTS
POST OFFICE BOX 655012 • DALLAS, TEXAS 75265

description

The '46A, '47A, and 'LS47 feature active-low outputs designed for driving common-anode LEDs or incandescent indicators directly. The '48, 'LS48, and 'LS49 feature active-high outputs for driving lamp buffers or common-cathode LEDs. All of the circuits except 'LS49 have full ripple-blanking input/output controls and a lamp test input. The 'LS49 circuit incorporates a direct blanking input. Segment identification and resultant displays are shown below. Display patterns for BCD input counts above 9 are unique symbols to authenticate input conditions.

The '46A, '47A, '48, 'LS47, and 'LS48 circuits incorporate automatic leading and/or trailing-edge zero-blanking control ($\overline{RBI}$ and $\overline{RBO}$). Lamp test ($\overline{LT}$) of these types may be performed at any time when the $\overline{BI}/\overline{RBO}$ node is at a high level. All types (including the '49 and 'LS49) contain an overriding blanking input ($\overline{BI}$), which can be used to control the lamp intensity by pulsing or to inhibit the outputs. Inputs and outputs are entirely compatible for use with TTL logic outputs.

The SN54246/SN74246 and '247 and the SN54LS247/SN74LS247 and 'LS248 compose the 6 and the 9 with tails and were designed to offer the designer a choice between two indicator fonts.

SEGMENT IDENTIFICATION

NUMERICAL DESIGNATIONS AND RESULTANT DISPLAYS

2

TTL Devices

'46A, '47A, 'LS47 FUNCTION TABLE (T1)

DECIMAL OR FUNCTION	INPUTS						$\overline{BI}/\overline{RBO}$†	OUTPUTS							NOTE
	$\overline{LT}$	$\overline{RBI}$	D	C	B	A		a	b	c	d	e	f	g	
0	H	H	L	L	L	L	H	ON	ON	ON	ON	ON	ON	OFF	
1	H	X	L	L	L	H	H	OFF	ON	ON	OFF	OFF	OFF	OFF	
2	H	X	L	L	H	L	H	ON	ON	OFF	ON	ON	OFF	ON	
3	H	X	L	L	H	H	H	ON	ON	ON	ON	OFF	OFF	ON	
4	H	X	L	H	L	L	H	OFF	ON	ON	OFF	OFF	ON	ON	
5	H	X	L	H	L	H	H	ON	OFF	ON	ON	OFF	ON	ON	
6	H	X	L	H	H	L	H	OFF	OFF	ON	ON	ON	ON	ON	
7	H	X	L	H	H	H	H	ON	ON	ON	OFF	OFF	OFF	OFF	1
8	H	X	H	L	L	L	H	ON	ON	ON	ON	ON	ON	ON	
9	H	X	H	L	L	H	H	ON	ON	ON	OFF	OFF	ON	ON	
10	H	X	H	L	H	L	H	OFF	OFF	OFF	ON	ON	OFF	ON	
11	H	X	H	L	H	H	H	OFF	OFF	ON	ON	OFF	OFF	ON	
12	H	X	H	H	L	L	H	OFF	ON	OFF	OFF	OFF	ON	ON	
13	H	X	H	H	L	H	H	ON	OFF	OFF	ON	OFF	ON	ON	
14	H	X	H	H	H	L	H	OFF	OFF	OFF	ON	ON	ON	ON	
15	H	X	H	H	H	H	H	OFF	OFF	OFF	OFF	OFF	OFF	OFF	
BI	X	X	X	X	X	X	L	OFF	OFF	OFF	OFF	OFF	OFF	OFF	2
RBI	H	L	L	L	L	L	L	OFF	OFF	OFF	OFF	OFF	OFF	OFF	3
LT	L	X	X	X	X	X	H	ON	ON	ON	ON	ON	ON	ON	4

H = high level, L = low level, X = irrelevant

NOTES 1. The blanking input ($\overline{BI}$) must be open or held at a high logic level when output functions 0 through 15 are desired. The ripple-blanking input ($\overline{RBI}$) must be open or high if blanking of a decimal zero is not desired.

2. When a low logic level is applied directly to the blanking input ($\overline{BI}$), all segment outputs are off regardless of the level of any other input.

3. When ripple-blanking input ($\overline{RBI}$) and inputs A, B, C, and D are at a low level with the lamp test input high, all segment outputs go off and the ripple-blanking output ($\overline{RBO}$) goes to a low level (response condition).

4. When the blanking input/ripple blanking output ($\overline{BI}/\overline{RBO}$) is open or held high and a low is applied to the lamp test input, all segment outputs are on.

†$\overline{BI}/\overline{RBO}$ is wire AND logic serving as blanking input ($\overline{BI}$) and/or ripple blanking output ($\overline{RBO}$).

SN5446A, '47A, '48, SN54LS47, 'LS48, 'LS49,
SN7446A, '47A, '48, SN74LS47, 'LS48, 'LS49
BCD-TO-SEVEN-SEGMENT DECODERS/DRIVERS

'48, 'LS48
FUNCTION TABLE (T2)

DECIMAL OR FUNCTION	INPUTS						BI/RBO†	OUTPUTS							NOTE
	LT	RBI	D	C	B	A		a	b	c	d	e	f	g	
0	H	H	L	L	L	L	H	H	H	H	H	H	H	L	
1	H	X	L	L	L	H	H	L	H	H	L	L	L	L	
2	H	X	L	L	H	L	H	H	H	L	H	H	L	H	
3	H	X	L	L	H	H	H	H	H	H	H	L	L	H	
4	H	X	L	H	L	L	H	L	H	H	L	L	H	H	
5	H	X	L	H	L	H	H	H	L	H	H	L	H	H	
6	H	X	L	H	H	L	H	L	L	H	H	H	H	H	
7	H	X	L	H	H	H	H	H	H	H	L	L	L	L	1
8	H	X	H	L	L	L	H	H	H	H	H	H	H	H	
9	H	X	H	L	L	H	H	H	H	H	L	L	H	H	
10	H	X	H	L	H	L	H	L	L	L	H	H	L	H	
11	H	X	H	L	H	H	H	L	L	H	H	L	L	H	
12	H	X	H	H	L	L	H	L	H	L	L	L	H	H	
13	H	X	H	H	L	H	H	H	L	L	H	L	H	H	
14	H	X	H	H	H	L	H	L	L	L	H	H	H	H	
15	H	X	H	H	H	H	H	L	L	L	L	L	L	L	
BI	X	X	X	X	X	X	L	L	L	L	L	L	L	L	2
RBI	H	L	L	L	L	L	L	L	L	L	L	L	L	L	3
LT	L	X	X	X	X	X	H	H	H	H	H	H	H	H	4

H = high level, L = low level, X = irrelevant

NOTES:
1. The blanking input (B̄Ī) must be open or held at a high logic level when output functions 0 through 15 are desired. The ripple-blanking input (R̄B̄Ī) must be open or high, if blanking of a decimal zero is not desired.
2. When a low logic level is applied directly to the blanking input (B̄Ī), all segment outputs are low regardless of the level of any other input.
3. When ripple-blanking input (R̄B̄Ī) and inputs A, B, C, and D are at a low level with the lamp-test input high, all segment outputs go low and the ripple-blanking output (R̄B̄O) goes to a low level (response condition).
4. When the blanking input/ripple-blanking output (B̄Ī/R̄B̄O) is open or held high and a low is applied to the lamp-test input, all segment outputs are high.

† B̄Ī/R̄B̄O is wire-AND logic serving as blanking input (B̄Ī) and/or ripple-blanking output (R̄B̄O).

'LS49
FUNCTION TABLE (T3)

DECIMAL OR FUNCTION	INPUTS					OUTPUTS							NOTE
	D	C	B	A	B̄Ī	a	b	c	d	e	f	g	
0	L	L	L	L	H	H	H	H	H	H	H	L	
1	L	L	L	H	H	L	H	H	L	L	L	L	
2	L	L	H	L	H	H	H	L	H	H	L	H	
3	L	L	H	H	H	H	H	H	H	L	L	H	
4	L	H	L	L	H	L	H	H	L	L	H	H	
5	L	H	L	H	H	H	L	H	H	L	H	H	
6	L	H	H	L	H	L	L	H	H	H	H	H	
7	L	H	H	H	H	H	H	H	L	L	L	L	1
8	H	L	L	L	H	H	H	H	H	H	H	H	
9	H	L	L	H	H	H	H	H	L	L	H	H	
10	H	L	H	L	H	L	L	L	H	H	L	H	
11	H	L	H	H	H	L	L	H	H	L	L	H	
12	H	H	L	L	H	L	H	L	L	L	H	H	
13	H	H	L	H	H	H	L	L	H	L	H	H	
14	H	H	H	L	H	L	L	L	H	H	H	H	
15	H	H	H	H	H	L	L	L	L	L	L	L	
BI	X	X	X	X	L	L	L	L	L	L	L	L	2

H = high level, L = low level, X = irrelevant

NOTES:
1. The blanking input (B̄Ī) must be open or held at a high logic level when output functions 0 through 15 are desired.
2. When a low logic level is applied directly to the blanking input (B̄Ī), all segment outputs are low regardless of the level of any other input.

TEXAS INSTRUMENTS

POST OFFICE BOX 655012 • DALLAS, TEXAS 75265

logic diagrams (positive logic)

'46A, '47A, 'LS47

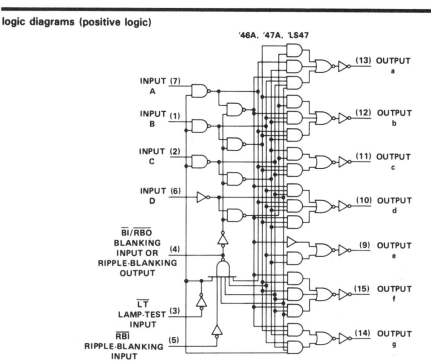

'48, 'LS48

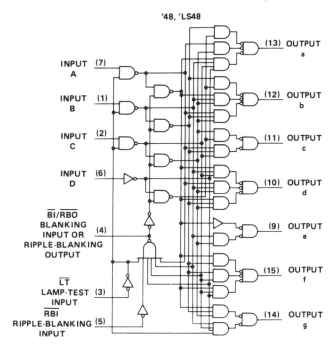

Pin numbers shown are for D, J, N, and W packages.

TEXAS
INSTRUMENTS
POST OFFICE BOX 655012 • DALLAS, TEXAS 75265

2-179

2

TTL Devices

logic diagrams (continued)

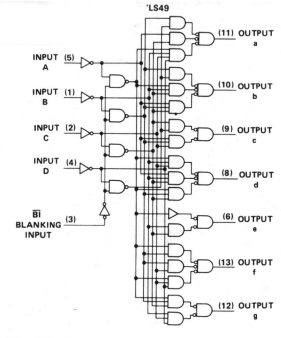

Pin numbers shown are for D, J, N, and W packages.

TEXAS
INSTRUMENTS
POST OFFICE BOX 655012 • DALLAS, TEXAS 75265

SN54LS47, 'LS48, 'LS49, SN74LS47, 'LS48, 'LS49
BCD-TO-SEVEN-SEGMENT DECODERS/DRIVERS

schematics of inputs and outputs

'LS47, 'LS48, 'LS49

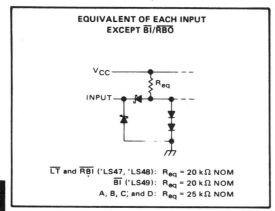

EQUIVALENT OF EACH INPUT
EXCEPT $\overline{BI}/\overline{RBO}$

$\overline{LT}$ and $\overline{RBI}$ ('LS47, 'LS48): R_{eq} = 20 kΩ NOM
$\overline{BI}$ ('LS49): R_{eq} = 20 kΩ NOM
A, B, C, and D: R_{eq} = 25 kΩ NOM

'LS47, 'LS48, 'LS49

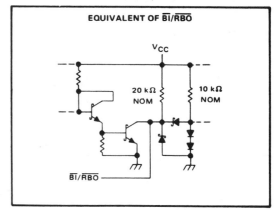

EQUIVALENT OF $\overline{BI}/\overline{RBO}$

'LS47

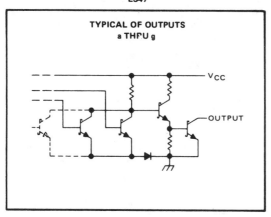

TYPICAL OF OUTPUTS
a THRU g

'LS48

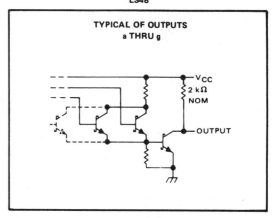

TYPICAL OF OUTPUTS
a THRU g

'LS49

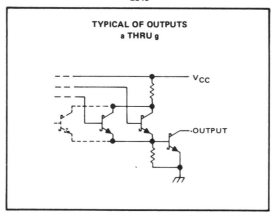

TYPICAL OF OUTPUTS
a THRU g

2

TTL Devices

absolute maximum ratings over operating free-air temperature range (unless otherwise noted)

Supply voltage, V_{CC} (see Note 1) . . . . 7 V
Input voltage . . . . 7 V
Peak output current ($t_w \leqslant 1$ ms, duty cycle $\leqslant 10\%$) . . . . 200 mA
Current forced into any output in the off state . . . . 1 mA
Operating free-air temperature range: SN54LS47 . . . . -55°C to 125°C
SN74LS47 . . . . 0°C to 70°C
Storage temperature range . . . . -65°C to 150°C

NOTE 1: Voltage values are with respect to network ground terminal.

recommended operating conditions

		SN54LS47 MIN	NOM	MAX	SN74LS47 MIN	NOM	MAX	UNIT
Supply voltage, V_{CC}		4.5	5	5.5	4.75	5	5.25	V
Off-state output voltage, $V_{O(off)}$	a thru g			15			15	V
On-state output current, $I_{O(on)}$	a thru g			12			24	mA
High-level output current, I_{OH}	$\overline{BI}/\overline{RBO}$			-50			-50	µA
Low-level output current, I_{OL}	$\overline{BI}/\overline{RBO}$			1.6			3.2	mA
Operating free-air temperature, T_A		-55		125	0		70	$^\circ$C

electrical characteristics over recommended operating free-air temperature range (unless otherwise noted)

PARAMETER		TEST CONDITIONS†		SN54LS47 MIN	TYP‡	MAX	SN74LS47 MIN	TYP‡	MAX	UNIT
V_{IH}	High-level input voltage			2			2			V
V_{IL}	Low-level input voltage					0.7			0.8	V
V_{IK}	Input clamp voltage	V_{CC} = MIN,	$I_I = -18$ mA			-1.5			-1.5	V
V_{OH}	High-level output voltage $\overline{BI}/\overline{RBO}$	V_{CC} = MIN, V_{IH} = 2 V, $V_{IL} = V_{IL}$ max, $I_{OH} = -50$ µA		2.4	4.2		2.4	4.2		V
V_{OL}	Low-level output voltage $\overline{BI}/\overline{RBO}$	V_{CC} = MIN, V_{IH} = 2 V, $V_{IL} = V_{IL}$ max	I_{OL} = 1.6 mA		0.25	0.4		0.25	0.4	V
			I_{OL} = 3.2 mA					0.35	0.5	
$I_{O(off)}$	Off-state output current a thru g	V_{CC} = MAX, V_{IH} = 2 V, $V_{IL} = V_{IL}$ max, $V_{O(off)}$ = 15 V				250			250	µA
$V_{O(on)}$	On-state output voltage a thru g	V_{CC} = MIN, V_{IH} = 2 V, $V_{IL} = V_{IL}$ max	$I_{O(on)}$ = 12 mA		0.25	0.4		0.25	0.4	V
			$I_{O(on)}$ = 24 mA					0.35	0.5	
I_I	Input current at maximum input voltage	V_{CC} = MAX, V_I = 7 V				0.1			0.1	mA
I_{IH}	High-level input current	V_{CC} = MAX, V_I = 2.7 V				20			20	µA
I_{IL}	Low-level input current	Any input except $\overline{BI}/\overline{RBO}$ V_{CC} = MAX, V_I = 0.4 V				-0.4			-0.4	mA
		$\overline{BI}/\overline{RBO}$				-1.2			-1.2	
I_{OS}	Short-circuit output current $\overline{BI}/\overline{RBO}$	V_{CC} = MAX		-0.3		-2	-0.3		-2	mA
I_{CC}	Supply current	V_{CC} = MAX,	See Note 2		7	13		7	13	mA

†For conditions shown as MIN or MAX, use the appropriate value specified under recommended operating conditions.
‡All typical values are at V_{CC} = 5 V, $T_A = 25^\circ$C.
NOTE 2: I_{CC} is measured with all outputs open and all inputs at 4.5 V.

switching characteristics, V_{CC} = 5 V, T_A = 25 °C

PARAMETER		TEST CONDITIONS	MIN	TYP	MAX	UNIT
t_{off}	Turn-off time from A input	C_L = 15 pF, R_L = 665 Ω, See Note 3			100	ns
t_{on}	Turn-on time from A input				100	
t_{off}	Turn-off time from $\overline{RBI}$ input, outputs (a-f) only				100	ns
t_{on}	Turn-on time from $\overline{RBI}$ input, outputs (a-f) only				100	

NOTE 3: Load circuits and voltage waveforms are shown in Section 1.

TEXAS
INSTRUMENTS

POST OFFICE BOX 655012 • DALLAS, TEXAS 75265

FUNCTION TABLE
(each latch)

INPUTS		OUTPUTS	
D	C	Q	$\overline{Q}$
L	H	L	H
H	H	H	L
X	L	Q_0	$\overline{Q_0}$

H = high level, L = low level, X = irrelevant

Q_0 = the level of Q before the high-to-low transition of G

description

These latches are ideally suited for use as temporary storage for binary information between processing units and input/output or indicator units. Information present at a data (D) input is transferred to the Q output when the enable (C) is high and the Q output will follow the data input as long as the enable remains high. When the enable goes low, the information (that was present at the data input at the time the transition occurred) is retained at the Q output until the enable is permitted to go high.

The '75 and 'LS75 feature complementary Q and $\overline{Q}$ outputs from a 4-bit latch, and are available in various 16-pin packages. For higher component density applications, the '77 and 'LS77 4-bit latches are available in 14-pin flat packages.

These circuits are completely compatible with all popular TTL families. All inputs are diode-clamped to minimize transmission-line effects and simplify system design. Series 54 and 54LS devices are characterized for operation over the full military temperature range of −55°C to 125°C; Series 74, and 74LS devices are characterized for operation from 0°C to 70°C.

SN5475, SN54LS75 . . . J OR W PACKAGE
SN7475 . . . N PACKAGE
SN74LS75 . . . D OR N PACKAGE
(TOP VIEW)

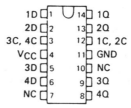

SN5477, SN54LS77 . . . W PACKAGE
(TOP VIEW)

NC - No internal connection

logic symbols[†]

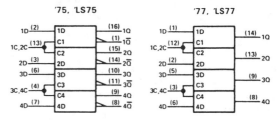

[†]These symbols are in accordance with ANSI/IEEE Std 91-1984 and IEC Publication 617-12.

absolute maximum ratings over operating free-air temperature range (unless otherwise noted)

Supply voltage, V_{CC} (See Note 1) . . . . 7 V
Input voltage: '75, '77 . . . . 5.5 V
           'LS75, 'LS77 . . . . 7 V
Interemitter voltage (see Note 2) . . . . 5.5 V
Operating free-air temperature range: SN54' . . . . −55°C to 125°C
                     SN74' . . . . 0°C to 70°C
Storage temperature range . . . . −65°C to 150°C

NOTES: 1. Voltage values are with respect to network ground terminal.
       2. This is the voltage between two emitters of a multiple-emitter input transistor and is not applicable to the 'LS75 and 'LS77.

TTL Devices 2

TEXAS
INSTRUMENTS

POST OFFICE BOX 655012 • DALLAS, TEXAS 75265

- Full-Carry Look-Ahead across the Four Bits
- Systems Achieve Partial Look-Ahead Performance with the Economy of Ripple Carry
- SN54283/SN74283 and SN54LS283/SN74LS283 Are Recommended For New Designs as They Feature Supply Voltage and Ground on Corner Pins to Simplify Board Layout

TYPE	TYPICAL ADD TIMES		TYPICAL POWER DISSIPATION PER 4-BIT ADDER
	TWO 8-BIT WORDS	TWO 16-BIT WORDS	
'83A	23 ns	43 ns	310 mW
'LS83A	25 ns	45 ns	95 mW

description

These improved full adders perform the addition of two 4-bit binary numbers. The sum (Σ) outputs are provided for each bit and the resultant carry (C4) is obtained from the fourth bit. These adders feature full internal look ahead across all four bits generating the carry term in ten nanoseconds typically. This provides the system designer with partial look-ahead performance at the economy and reduced package count of a ripple-carry implementation.

The adder logic, including the carry, is implemented in its true form meaning that the end-around carry can be accomplished without the need for logic or level inversion.

Designed for medium-speed applications, the circuits utilize transistor-transistor logic that is compatible with most other TTL families and other saturated low-level logic families.

Series 54 and 54LS circuits are characterized for operation over the full military temperature range of $-55°C$ to $125°C$, and Series 74 and 74LS circuits are characterized for operation from $0°C$ to $70°C$.

logic symbol[†]

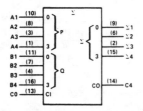

[†]This symbol is in accordance with ANSI/IEEE Std 91-1984 and IEC Publication 617-12.
Pin numbers are for D, J, N, and W packages.

SN5483A,SN54LS83A . . . J OR W PACKAGE
SN7483A . . . N PACKAGE
SN74LS83A . . . D OR N PACKAGE
(TOP VIEW)

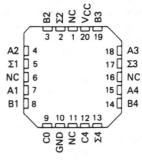

SN54LS83A . . . FK PACKAGE
(TOP VIEW)

NC - No internal connection

FUNCTION TABLE

INPUT				OUTPUT					
				WHEN C0 = L			WHEN C0 = H		
						WHEN C2 = L			WHEN C2 = H
A1 / A3	B1 / B3	A2 / A4	B2 / B4	Σ1 / Σ3	Σ2 / Σ4	C2 / C4	Σ1 / Σ3	Σ2 / Σ4	C2 / C4
L	L	L	L	L	L	L	H	L	L
H	L	L	L	H	L	L	L	H	L
L	H	L	L	H	L	L	L	H	L
H	H	L	L	L	H	L	H	H	L
L	L	H	L	L	H	L	H	H	L
H	L	H	L	H	H	L	L	L	H
L	H	H	L	H	H	L	L	L	H
H	H	H	L	L	L	H	H	H	L
L	L	L	H	L	H	L	L	H	L
H	L	L	H	H	H	L	L	L	H
L	H	L	H	H	H	L	L	L	H
H	H	L	H	L	L	H	H	H	L
L	L	H	H	L	L	H	H	L	L
H	L	H	H	H	L	H	L	H	H
L	H	H	H	H	L	H	L	H	H
H	H	H	H	L	H	H	H	H	H

H = high level, L = low level

NOTE: Input conditions at A1, B1, A2, B2, and C0 are used to determine outputs Σ1 and Σ2 and the value of the internal carry C2. The values at C2, A3, B3, A4, and B4 are then used to determine outputs Σ3, Σ4, and C4.

TEXAS INSTRUMENTS
POST OFFICE BOX 655012 • DALLAS. TEXAS 75265

2

TTL Devices

logic diagram (positive logic)

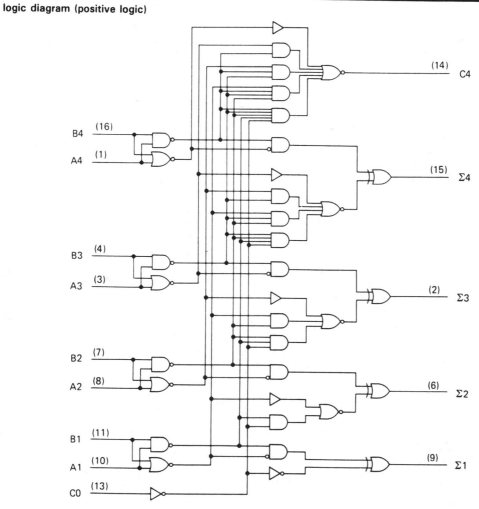

2

TTL Devices

Pin numbers shown are for D, J, N, and W packages.

absolute maximum ratings over operating free-air temperature range (unless otherwise noted)

Supply voltage, V_{CC} (see Note 1)	7 V
Input voltage: '83A	5.5 V
'LS83A	7 V
Interemitter voltage (see Note 2)	5.5 V
Operating free-air temperature range: SN5483A, SN54LS83A	−55°C to 125°C
SN7483A, SN74LS83A	0°C to 70°C
Storage temperature range	−65°C to 150°C

NOTES: 1. Voltage values, except interemitter voltage, are with respect to network ground terminal.
2. This is the voltage between two emitters of a multiple-emitter transistor. This rating applies for the '83A only between the following pairs: A1 and B1, A2 and B2, A3 and B3, A4 and B4.

**TEXAS
INSTRUMENTS**
POST OFFICE BOX 555012 • DALLAS, TEXAS 75265

TYPE	TYPICAL POWER DISSIPATION	TYPICAL DELAY (4-BIT WORDS)
'85	275 mW	23 ns
'LS85	52 mW	24 ns
'S85	365 mW	11 ns

description

These four-bit magnitude comparators perform comparison of straight binary and straight BCD (8-4-2-1) codes. Three fully decoded decisions about two 4-bit words (A, B) are made and are externally available at three outputs. These devices are fully expandable to any number of bits without external gates. Words of greater length may be compared by connecting comparators in cascade. The A > B, A < B, and A = B outputs of a stage handling less-significant bits are connected to the corresponding A > B, A*< B, and A = B inputs of the next stage handling more-significant bits. The stage handling the least-significant bits must have a high-level voltage applied to the A = B input. The cascading paths of the '85, 'LS85, and 'S85 are implemented with only a two-gate-level delay to reduce overall comparison times for long words. An alternate method of cascading which further reduces the comparison time is shown in the typical application data.

SN5485, SN54LS85, SN54S85 . . . J OR W PACKAGE
SN7485 . . . N PACKAGE
SN74LS85, SN74S85 . . . D OR N PACKAGE
(TOP VIEW)

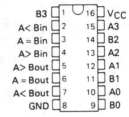

SN54LS85, SN54S85 . . . FK PACKAGE
(TOP VIEW)

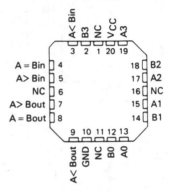

NC · No internal connection

FUNCTION TABLE

COMPARING INPUTS				CASCADING INPUTS			OUTPUTS		
A3, B3	A2, B2	A1, B1	A0, B0	A > B	A < B	A = B	A > B	A < B	A = B
A3 > B3	X	X	X	X	X	X	H	L	L
A3 < B3	X	X	X	X	X	X	L	H	L
A3 = B3	A2 > B2	X	X	X	X	X	H	L	L
A3 = B3	A2 < B2	X	X	X	X	X	L	H	L
A3 = B2	A2 = B2	A1 > B1	X	X	X	X	H	L	L
A3 = B3	A2 = B2	A1 < B1	X	X	X	X	L	H	L
A2 = B3	A2 = B2	A1 = B1	A0 > B0	X	X	X	H	L	L
A3 = B3	A2 = B2	A1 = B1	A0 < B0	X	X	X	L	H	L
A3 = B3	A2 = B2	A1 = B1	A0 = B0	H	L	L	H	L	L
A3 = B3	A2 = B2	A1 = B1	A0 = B0	L	H	L	L	H	L
A3 = B3	A2 = B2	A1 = B1	A0 = B0	X	X	H	L	L	H
A3 = B3	A2 = B2	A1 = B1	A0 = B0	H	H	L	L	L	L
A3 = B3	A2 = B2	A1 = B1	A0 = B0	L	L	L	H	H	L

TEXAS
INSTRUMENTS
POST OFFICE BOX 655012 • DALLAS, TEXAS 75265

TTL Devices **2**

SN5485, SN54LS85, SN54S85,
SN7485, SN74LS85, SN74S85
4-BIT MAGNITUDE COMPARATORS

logic diagrams (positive logic)

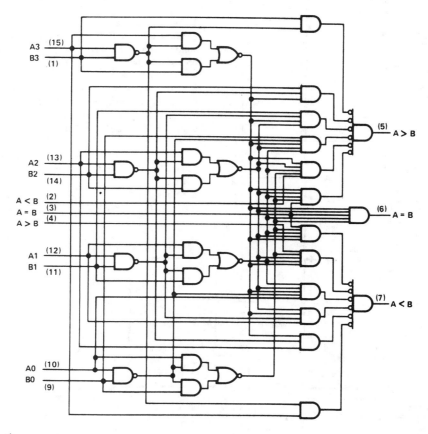

logic symbol†

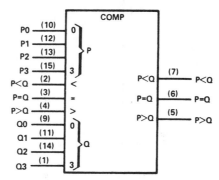

†This symbol is in accordancae with ANSI/IEEE Std 91-1984 and IEC Publication 617-12.
Pin numbers shown are for D, J, N, and W packages.

TEXAS
INSTRUMENTS
POST OFFICE BOX 655012 • DALLAS, TEXAS 75265

TYPICAL APPLICATION DATA

COMPARISON OF TWO N-BIT WORDS

This application demonstrates how these magnitude comparators can be cascaded to compare longer words. The example illustrated shows the comparison of two 24-bit words; however, the design is expandable to n-bits. As an example, one comparator can be used with five of the 24-bit comparators illustrated to expand the word length to 120-bits. Typical comparison times for various word lengths using the '85, 'LS85, or 'S85 are:

WORD LENGTH	NUMBER OF PKGS	'85	'LS85	'S85
1-4 bits	1	23 ns	24 ns	11 ns
5-24 bits	2-6	46 ns	48 ns	22 ns
25-120 bits	8-31	69 ns	72 ns	33 ns

2

TTL Devices

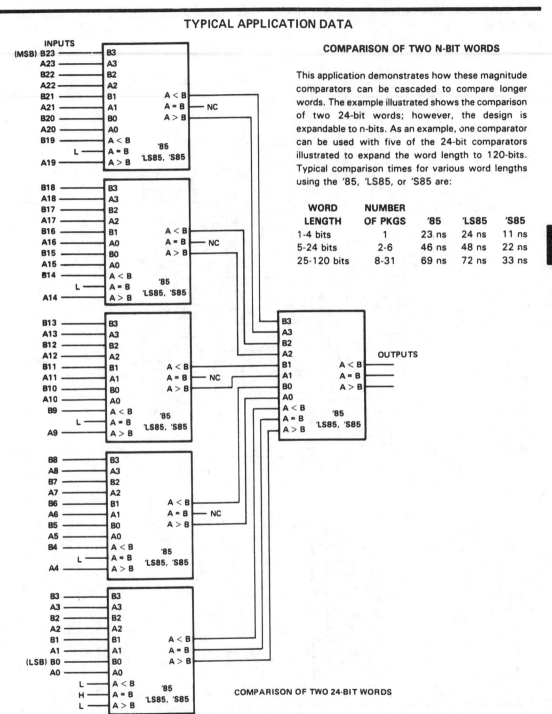

COMPARISON OF TWO 24-BIT WORDS

TEXAS INSTRUMENTS
POST OFFICE BOX 655012 • DALLAS, TEXAS 75265

2-269

- Package Options Include Plastic "Small Outline" Packages, Ceramic Chip Carriers and Flat Packages, and Standard Plastic and Ceramic 300-mil DIPs

- Dependable Texas Instruments Quality and Reliability

TYPE	TYPICAL AVERAGE PROPAGATION DELAY TIME	TYPICAL TOTAL POWER DISSIPATION
'86	14 ns	150 mW
'LS86A	10 ns	30.5 mW
'S86	7 ns	250 mW

SN5486, SN54LS86A, SN54S86 . . . J OR W PACKAGE
SN7486 . . . N PACKAGE
SN74LS86A, SN74S86 . . . D OR N PACKAGE
(TOP VIEW)

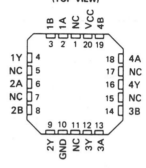

```
1A  [1   U  14] Vcc
1B  [2      13] 4B
1Y  [3      12] 4A
2A  [4      11] 4Y
2B  [5      10] 3B
2Y  [6       9] 3A
GND [7       8] 3Y
```

SN54LS86A, SN54S86 . . . FK PACKAGE
(TOP VIEW)

```
        1B 1A NC Vcc 4B
         3  2  1  20 19
    1Y [ 4          18 ] 4A
    NC [ 5          17 ] NC
    2A [ 6          16 ] 4Y
    NC [ 7          15 ] NC
    2B [ 8          14 ] 3B
         9 10 11 12 13
        2Y GND NC 3Y 3A
```

NC - No internal connection

description

These devices contain four independent 2-input Exclusive-OR gates. They perform the Boolean functions $Y = A \oplus B = \bar{A}B + A\bar{B}$ in positive logic.

A common application is as a true/complement element. If one of the inputs is low, the other input will be reproduced in true form at the output. If one of the inputs is high, the signal on the other input will be reproduced inverted at the output.

The SN5486, 54LS86A, and the SN54S86 are characterized for operation over the full military temperature range of −55°C to 125°C. The SN7486, SN74LS86A, and the SN74S86 are characterized for operation from 0°C to 70°C.

exclusive-OR logic

An exclusive-OR gate has many applications, some of which can be represented better by alternative logic symbols.

EXCLUSIVE-OR

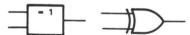 

These are five equivalent Exclusive-OR symbols valid for an '86 or 'LS86A gate in positive logic; negation may be shown at any two ports.

LOGIC IDENTITY ELEMENT	EVEN-PARITY	ODD-PARITY ELEMENT
		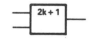

The output is active (low) if all inputs stand at the same logic level (i.e., A = B).

The output is active (low) if an even number of inputs (i.e., 0 or 2) are active.

The output is active (high) if an odd number of inputs (i.e., only 1 of the 2) are active.

TEXAS INSTRUMENTS
POST OFFICE BOX 655012 • DALLAS, TEXAS 75265

TTL Devices

2

SN5486, SN54LS86A, SN54S86,
SN7486, SN74LS86A, SN74S86
QUADRUPLE 2-INPUT EXCLUSIVE-OR GATES

schematics of inputs and outputs

'86

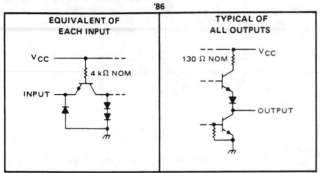

EQUIVALENT OF EACH INPUT	TYPICAL OF ALL OUTPUTS

'LS86A

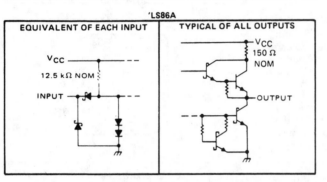

EQUIVALENT OF EACH INPUT	TYPICAL OF ALL OUTPUTS

'S86

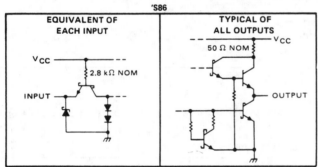

EQUIVALENT OF EACH INPUT	TYPICAL OF ALL OUTPUTS

logic symbol†

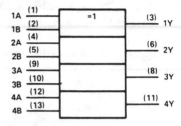

†This symbol is in accordance with
ANSI/IEEE Std. 91-1984 and IEC Publication 617-12.
Pin numbers shown are for D, J, N, and W packages.

FUNCTION TABLE

INPUTS		OUTPUT
A	B	Y
L	L	L
L	H	H
H	L	H
H	H	L

H = high level, L = low level

TEXAS INSTRUMENTS
POST OFFICE BOX 655012 • DALLAS, TEXAS 75265

'90A, 'LS90 . . . Decade Counters

'92A, 'LS92 . . . Divide By-Twelve Counters

'93A, 'LS93 . . . 4-Bit Binary Counters

TYPES	TYPICAL POWER DISSIPATION
'90A	145 mW
'92A, '93A	130 mW
'LS90, 'LS92, 'LS93	45 mW

description

Each of these monolithic counters contains four master-slave flip-flops and additional gating to provide a divide-by-two counter and a three-stage binary counter for which the count cycle length is divide-by-five for the '90A and 'LS90, divide-by-six for the '92A and 'LS92, and the divide-by-eight for the '93A and 'LS93.

All of these counters have a gated zero reset and the '90A and 'LS90 also have gated set-to-nine inputs for use in BCD nine's complement applications.

To use their maximum count length (decade, divide-by-twelve, or four-bit binary) of these counters, the CKB input is connected to the Q_A output. The input count pulses are applied to CKA input and the outputs are as described in the appropriate function table. A symmetrical divide-by-ten count can be obtained from the '90A or 'LS90 counters by connecting the Q_D output to the CKA input and applying the input count to the CKB input which gives a divide-by-ten square wave at output Q_A.

SN5490A, SN54LS90 . . . J OR W PACKAGE
SN7490A . . . N PACKAGE
SN74LS90 . . . D OR N PACKAGE
(TOP VIEW)

```
         _____
CKB  [1       14]  CKA
R0(1)[2       13]  NC
R0(2)[3       12]  QA
NC   [4       11]  QD
VCC  [5       10]  GND
R9(1)[6        9]  QB
R9(2)[7        8]  QC
```

SN5492A, SN54LS92 . . . J OR W PACKAGE
SN7492A . . . N PACKAGE
SN74LS92 . . . D OR N PACKAGE
(TOP VIEW)

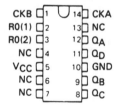

```
         _____
CKB  [1       14]  CKA
NC   [2       13]  NC
NC   [3       12]  QA
NC   [4       11]  QB
VCC  [5       10]  GND
R0(1)[6        9]  QC
R0(2)[7        8]  QD
```

SN5493A, SN54LS93 . . . J OR W PACKAGE
SN7493 . . . N PACKAGE
SN74LS93 . . . D OR N PACKAGE
(TOP VIEW)

```
         _____
CKB  [1       14]  CKA
R0(1)[2       13]  NC
R0(2)[3       12]  QA
NC   [4       11]  QD
VCC  [5       10]  GND
NC   [6        9]  QB
NC   [7        8]  QC
```

NC—No internal connection

TEXAS
INSTRUMENTS
POST OFFICE BOX 655012 • DALLAS, TEXAS 75265

TTL Devices

2

SN5490A, '92A, '93A, SN54LS90, 'LS92, 'LS93, SN7490A, '92A, '93A, SN74LS90, 'LS92, 'LS93
DECADE, DIVIDE-BY-TWELVE, AND BINARY COUNTERS

logic symbols[†]

'90

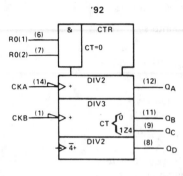

'92

'93A, 'LS93

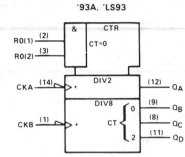

[†]These symbols are in accordance with ANSI/IEEE Std. 91-1984 and IEC Publication 617-12.

TEXAS
INSTRUMENTS
POST OFFICE BOX 655012 • DALLAS, TEXAS 75265

'90A, 'LS90
BCD COUNT SEQUENCE
(See Note A)

COUNT	OUTPUT			
	Q_D	Q_C	Q_B	Q_A
0	L	L	L	L
1	L	L	L	H
2	L	L	H	L
3	L	L	H	H
4	L	H	L	L
5	L	H	L	H
6	L	H	H	L
7	L	H	H	H
8	H	L	L	L
9	H	L	L	H

'90A, 'LS90
BI-QUINARY (5-2)
(See Note B)

COUNT	OUTPUT			
	Q_A	Q_D	Q_C	Q_B
0	L	L	L	L
1	L	L	L	H
2	L	L	H	L
3	L	L	H	H
4	L	H	L	L
5	H	L	L	L
6	H	L	L	H
7	H	L	H	L
8	H	L	H	H
9	H	H	L	L

'92A, 'LS92
COUNT SEQUENCE
(See Note C)

COUNT	OUTPUT			
	Q_D	Q_C	Q_B	Q_A
0	L	L	L	L
1	L	L	L	H
2	L	L	H	L
3	L	L	H	H
4	L	H	L	L
5	L	H	L	H
6	H	L	L	L
7	H	L	L	H
8	H	L	H	L
9	H	L	H	H
10	H	H	L	L
11	H	H	L	H

'90A, 'LS90
RESET/COUNT FUNCTION TABLE

RESET INPUTS				OUTPUT			
$R_{0(1)}$	$R_{0(2)}$	$R_{9(1)}$	$R_{9(2)}$	Q_D	Q_C	Q_B	Q_A
H	H	L	X	L	L	L	L
H	H	X	L	L	L	L	L
X	X	H	H	H	L	L	H
X	L	X	L	COUNT			
L	X	L	X	COUNT			
L	X	X	L	COUNT			
X	L	L	X	COUNT			

'93A, 'LS93
COUNT SEQUENCE
(See Note C)

COUNT	OUTPUT			
	Q_D	Q_C	Q_B	Q_A
0	L	L	L	L
1	L	L	L	H
2	L	L	H	L
3	L	L	H	H
4	L	H	L	L
5	L	H	L	H
6	L	H	H	L
7	L	H	H	H
8	H	L	L	L
9	H	L	L	H
10	H	L	H	L
11	H	L	H	H
12	H	H	L	L
13	H	H	L	H
14	H	H	H	L
15	H	H	H	H

'92A, 'LS92, '93A, 'LS93
RESET/COUNT FUNCTION TABLE

RESET INPUTS		OUTPUT			
$R_{0(1)}$	$R_{0(2)}$	Q_D	Q_C	Q_B	Q_A
H	H	L	L	L	L
L	X	COUNT			
X	L	COUNT			

NOTES: A. Output Q_A is connected to input CKB for BCD count.
B. Output Q_D is connected to input CKA for bi-quinary count.
C. Output Q_A is connected to input CKB.
D. H = high level, L = low level, X = irrelevant

2

TTL Devices

SN5490A, '92A, '93A, SN54LS90, 'LS92, 'LS93, SN7490A, '92A, '93A, SN74LS90, 'LS92, 'LS93 DECADE, DIVIDE-BY-TWELVE, AND BINARY COUNTERS

logic diagrams (positive logic)

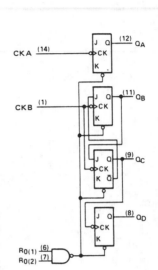

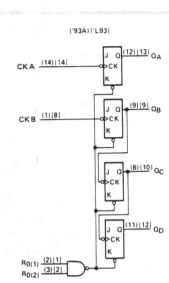

The J and K inputs shown without connection are for reference only and are functionally at a high level.

Pin numbers shown in ( ) are for the 'LS93 and '93A and pin numbers shown in [ ] are for the 54L93.

schematics of inputs and outputs

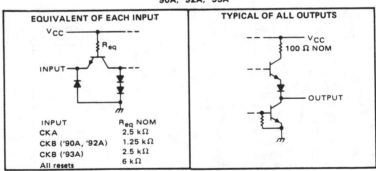

'90A, '92A, '93A

EQUIVALENT OF EACH INPUT	TYPICAL OF ALL OUTPUTS

INPUT	R_{eq} NOM
CKA	2.5 kΩ
CKB ('90A, '92A)	1.25 kΩ
CKB ('93A)	2.5 kΩ
All resets	6 kΩ

TEXAS
INSTRUMENTS
POST OFFICE BOX 655012 • DALLAS, TEXAS 75265

TYPE	TYPICAL MAXIMUM CLOCK FREQUENCY	TYPICAL POWER DISSIPATION
'95A	36 MHz	195 mW
'LS95B	36 MHz	65 mW

description

These 4-bit registers feature parallel and serial inputs, parallel outputs, mode control, and two clock inputs. The registers have three modes of operation:

Parallel (broadside) load
Shift right (the direction Q_A toward Q_D)
Shift left (the direction Q_D toward Q_A)

Parallel loading is accomplished by applying the four bits of data and taking the mode control input high. The data is loaded into the associated flip-flops and appears at the outputs after the high-to-low transition of the clock-2 input. During loading, the entry of serial data is inhibited.

Shift right is accomplished on the high-to-low transition of clock 1 when the mode control is low; shift left is accomplished on the high-to-low transition of clock 2 when the mode control is high by connecting the output of each flip-flop to the parallel input of the previous flip-flop (Q_D to input C, etc.) and serial data is entered at input D. The clock input may be applied commonly to clock 1 and clock 2 if both modes can be clocked from the same source. Changes at the mode control input should normally be made while both clock inputs are low; however, conditions described in the last three lines of the function table will also ensure that register contents are protected.

SN5495A, SN54LS95B . . . J OR W PACKAGE
SN7495A . . . N PACKAGE
SN74LS95B . . . D OR N PACKAGE
(TOP VIEW)

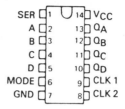

```
SER  [ 1    14 ] VCC
  A  [ 2    13 ] QA
  B  [ 3    12 ] QB
  C  [ 4    11 ] QC
  D  [ 5    10 ] QD
MODE [ 6     9 ] CLK 1
 GND [ 7     8 ] CLK 2
```

SN54LS95B . . . FK PACKAGE
(TOP VIEW)

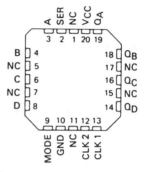

NC - No internal connection

2

TTL Devices

FUNCTION TABLE

MODE CONTROL	CLOCKS 2 (L)	CLOCKS 1 (R)	SERIAL	PARALLEL A	B	C	D	Q_A	Q_B	Q_C	Q_D
H	H	X	X	X	X	X	X	Q_{A0}	Q_{B0}	Q_{C0}	Q_{D0}
H	↓	X	X	a	b	c	d	a	b	c	d
H	↓	X	X	Q_B↑	Q_C↑	Q_D↑	d	Q_{Bn}	Q_{Cn}	Q_{Dn}	d
L	L	H	X	X	X	X	X	Q_{A0}	Q_{B0}	Q_{C0}	Q_{D0}
L	X	↓	H	X	X	X	X	H	Q_{An}	Q_{Bn}	Q_{Cn}
L	X	↓	L	X	X	X	X	L	Q_{An}	Q_{Bn}	Q_{Cn}
↑	L	L	X	X	X	X	X	Q_{A0}	Q_{B0}	Q_{C0}	Q_{D0}
↓	L	L	X	X	X	X	X	Q_{A0}	Q_{B0}	Q_{C0}	Q_{D0}
↓	L	H	X	X	X	X	X	Q_{A0}	Q_{B0}	Q_{C0}	Q_{D0}
↑	H	L	X	X	X	X	X	Q_{A0}	Q_{B0}	Q_{C0}	Q_{D0}
↑	H	H	X	X	X	X	X	Q_{A0}	Q_{B0}	Q_{C0}	Q_{D0}

↑Shifting left requires external connection of Q_B to A, Q_C to B, and Q_D to C. Serial data is entered at input D.
H = high level (steady state), L = low level (steady state), X = irrelevant (any input, including transitions)
↓ = transition from high to low level, ↑ = transition from low to high level
a, b, c, d = the level of steady state input at inputs A, B, C, or D, respectively.
Q_{A0}, Q_{B0}, Q_{C0}, Q_{D0} = the level of Q_A, Q_B, Q_C, or Q_D, respectively, before the indicated steady state input conditions were established.
Q_{An}, Q_{Bn}, Q_{Cn}, Q_{Dn} = the level of Q_A, Q_B, Q_C, or Q_D, respectively, before the most recent ↓ transition of the clock.

TEXAS
INSTRUMENTS

POST OFFICE BOX 655012 • DALLAS, TEXAS 75265

SN5495A, SN54LS95B, SN7495A, SN74LS95B
4-BIT PARALLEL-ACCESS SHIFT REGISTERS

logic symbol†

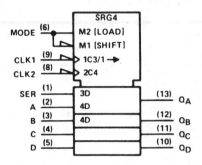

† This symbol is in accordance with ANSI/IEEE Std 91-1984 and IEC Publication 617-12.
Pin numbers shown are for D, J, N, and W packages.

logic diagram (positive logic)

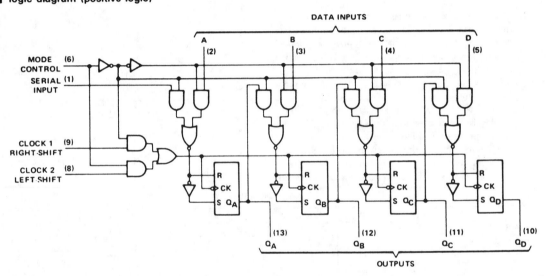

TEXAS
INSTRUMENTS

POST OFFICE BOX 655012 • DALLAS, TEXAS 75265

SN54LS112A, SN54S112, SN74LS112A, SN74S112A
DUAL J-K NEGATIVE-EDGE-TRIGGERED FLIP-FLOPS WITH PRESET AND CLEAR

D2661, APRIL 1982 – REVISED MARCH 1988

- **Fully Buffered to Offer Maximum Isolation from External Disturbance**
- **Package Options Include Plastic "Small Outline" Packages, Ceramic Chip Carriers and Flat Packages, and Plastic and Ceramic DIPs**
- **Dependable Texas Instruments Quality and Reliability**

description

These devices contain two independent J-K negative-edge-triggered flip-flops. A low level at the preset and clear inputs sets or resets the outputs regardless of the levels of the other inputs. When preset and clear are inactive (high), data at the J and K inputs meeting the setup time requirements are transferred to the outputs on the negative-going edge of the clock pulse. Clock triggering occurs at a voltage level and is not directly related to the rise time of the clock pulse. Following the hold time interval, data at the J and K inputs may be changed without affecting the levels at the outputs. These versatile flip-flops can perform as toggle flip-flops by tying J and K high.

The SN54LS112A and SN54S112 are characterized for operation over the full military temperature range of −55°C to 125°C. The SN74LS112A and SN74S112A are characterized for operation from 0°C to 70°C.

SN54LS112A, SN54S112 . . . J OR W PACKAGE
SN74LS112A, SN74S112A . . . D OR N PACKAGE
(TOP VIEW)

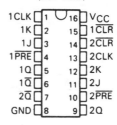

SN54LS112A, SN54S112 . . . FK PACKAGE
(TOP VIEW)

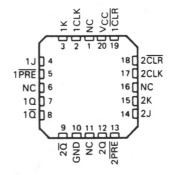

NC – No internal connection

logic symbol‡

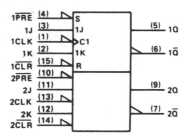

‡This symbol is in accordance with ANSI/IEEE Std 91-1984 and IEC Publication 617-12.

Pin numbers shown are for D, J, N, and W packages.

FUNCTION TABLE (each flip-flop)

INPUTS					OUTPUTS	
PRE	CLR	CLK	J	K	Q	Q̄
L	H	X	X	X	H	L
H	L	X	X	X	L	H
L	L	X	X	X	H†	H†
H	H	↓	L	L	Q₀	Q̄₀
H	H	↓	H	L	H	L
H	H	↓	L	H	L	H
H	H	↓	H	H	TOGGLE	
H	H	H	X	X	Q₀	Q̄₀

† The output levels in this configuration are not guaranteed to meet the minimum levels for V_{OH} if the lows at preset and clear are near V_{IL} minimum. Furthermore, this configuration is nonstable; that is, it will not persist when either preset or clear returns to its inactive (high) level.

TEXAS INSTRUMENTS
POST OFFICE BOX 655012 • DALLAS, TEXAS 75265

Copyright © 1982, Texas Instruments Incorporated

TTL Devices

2

SN54LS112A, SN54S112, SN74LS112A, SN74S112A
DUAL J-K NEGATIVE-EDGE-TRIGGERED
FLIP-FLOPS WITH PRESET AND CLEAR

logic diagrams (positive logic)

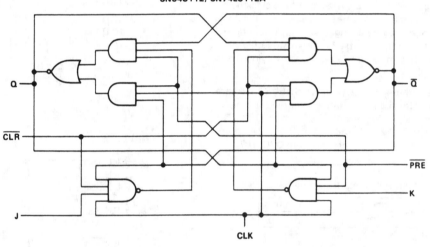

'LS112A

SN54S112, SN74LS112A

schematics of inputs and outputs

'LS112A

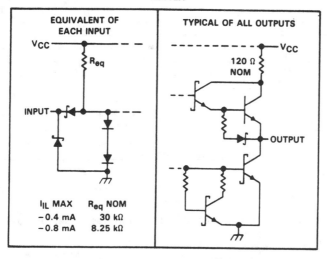

SN54S112, SN74S112A

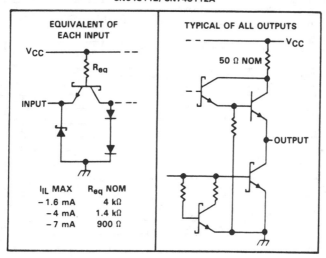

absolute maximum ratings over operating free-air temperature range (unless otherwise noted)

Supply voltage, V_{CC} (see Note 1) . . . . 7 V
Input voltage: 'LS112A . . . . 7 V
               SN54LS112, SN74LS112A . . . . 5.5 V
Operating free-air temperature range: SN54' . . . . −55°C to 125°C
                       SN74' . . . . 0°C to 70°C
Storage temperature range . . . . −65°C to 150°C

NOTE 1: Voltage values are with respect to network ground terminal.

TTL Devices

2

SN54LS112A, SN74LS112A
DUAL J-K NEGATIVE-EDGE-TRIGGERED
FLIP-FLOPS WITH PRESET AND CLEAR

recommended operating conditions

			SN54LS112A			SN74LS112A			UNIT
			MIN	NOM	MAX	MIN	NOM	MAX	
V_{CC}	Supply voltage		4.5	5	5.5	4.75	5	5.25	V
V_{IH}	High-level input voltage		2			2			V
V_{IL}	Low-level input voltage				0.7			0.8	V
I_{OH}	High-level output current				−0.4			−0.4	mA
I_{OL}	Low-level output current				4			8	mA
f_{clock}	Clock frequency		0		30	0		30	MHz
t_w	Pulse duration	CLK high	20			20			ns
		$\overline{PRE}$ or $\overline{CLR}$ low	25			25			
t_{su}	Set up time-before CLK↓	Data high or low	20			20			ns
		$\overline{CLR}$ inactive	25			25			
		$\overline{PRE}$ inactive	20			20			
t_h	Hold time-data after CLK↓		0			0			ns
T_A	Operating free-air temperature		−55		125	0		70	°C

electrical characteristics over recommended operating free-air temperature range (unless otherwise noted)

PARAMETER		TEST CONDITIONS†			SN54LS112A			SN74LS112A			UNIT
					MIN	TYP‡	MAX	MIN	TYP‡	MAX	
V_{IK}		V_{CC} = MIN,	I_I = −18 mA				−1.5			−1.5	V
V_{OH}		V_{CC} = MIN, I_{OH} = −0.4 mA	V_{IH} = 2 V,	V_{IL} = MAX,	2.5	3.4		2.7	3.4		V
V_{OL}		V_{CC} = MIN, I_{OL} = 4 mA	V_{IL} = MAX,	V_{IH} = 2 V,		0.25	0.4		0.25	0.4	V
		V_{CC} = MIN, I_{OL} = 8 mA	V_{IL} = MAX,	V_{IH} = 2 V,					0.35	0.5	
I_I	J or K	V_{CC} = MAX,	V_I = 7 V				0.1			0.1	mA
	$\overline{CLR}$ or $\overline{PRE}$						0.3			0.3	
	CLK						0.4			0.4	
I_{IH}	J or K	V_{CC} = MAX,	V_I = 2.7 V				20			20	μA
	$\overline{CLR}$ or $\overline{PRE}$						60			60	
	CLK						80			80	
I_{IL}	J or K	V_{CC} = MAX,	V_I = 0.4 V				−0.4			−0.4	mA
	All other						−0.8			−0.8	
I_{OS}§		V_{CC} = MAX,	see Note 2		−20		−100	−20		−100	mA
I_{CC} (Total)		V_{CC} = MAX,	see Note 3			4	6		4	6	mA

† For conditions shown as MIN or MAX, use the appropriate value specified under recommended operating conditions.
‡ All typical values are at V_{CC} = 5 V, T_A = 25°C.
§ Not more than one output should be shorted at a time, and the duration of the short-circuit should not exceed one second.

NOTES: 2. For certain devices where state commutation can be caused by shorting an output to ground, an equivalent test may be performed with V_O = 2.25 V and 2.125 V for the '54 family and the '74 family, respectively, with the minimum and maximum limits reduced to one half of their stated values.
3. With all outputs open, I_{CC} is measured with the Q and $\overline{Q}$ outputs high in turn. At the time of measurement, the clock input is grounded.

TEXAS INSTRUMENTS
POST OFFICE BOX 655012 • DALLAS, TEXAS 75265

2

TTL Devices

switching characteristics, V_{CC} = 5 V, T_A = 25°C (see Note 4)

PARAMETER	FROM (INPUT)	TO (OUTPUT)	TEST CONDITIONS	MIN	TYP	MAX	UNIT
f_{max}			R_L = 2 kΩ,　　C_L = 15 pF	30	45		MHz
t_{PLH}	$\overline{CLR}$, $\overline{PRE}$ or CLK	Q or $\overline{Q}$			15	20	ns
t_{PHL}					15	20	ns

NOTE 4: Load circuits and voltage waveforms are shown in Section 1.

2

TTL Devices

TEXAS
INSTRUMENTS

POST OFFICE BOX 655012 • DALLAS, TEXAS 75265

- **Designed Specifically for High-Speed:**
  Memory Decoders
  Data Transmission Systems

- **3 Enable Inputs to Simplify Cascading and/or Data Reception**

- **Schottky-Clamped for High Performance**

description

These Schottky-clamped TTL MSI circuits are designed to be used in high-performance memory decoding or data-routing applications requiring very short propagation delay times. In high-performance memory systems, these docoders can be used to minimize the effects of system decoding. When employed with high-speed memories utilizing a fast enable circuit, the delay times of these decoders and the enable time of the memory are usually less than the typical access time of the memory. This means that the effective system delay introduced by the Schottky-clamped system decoder is negligible.

The 'LS138, SN54S138, and SN74S138A decode one of eight lines dependent on the conditions at the three binary select inputs and the three enable inputs. Two active-low and one active-high enable inputs reduce the need for external gates or inverters when expanding. A 24-line decoder can be implemented without external inverters and a 32-line decoder requires only one inverter. An enable input can be used as a data input for demultiplexing applications.

All of these decoder/demultiplexers feature fully buffered inputs, each of which represents only one normalized load to its driving circuit. All inputs are clamped with high-performance Schottky diodes to suppress line-ringing and to simplify system design.

The SN54LS138 and SN54S138 are characterized for operation over the full military temperature range of −55°C to 125°C. The SN74LS138 and SN74S138A are characterized for operation from 0°C to 70°C.

SN54LS138, SN54S138 . . . J OR W PACKAGE
SN74LS138, SN74S138A . . . D OR N PACKAGE
(TOP VIEW)

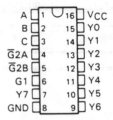

A	1	16	V_{CC}
B	2	15	Y0
C	3	14	Y1
$\overline{G2A}$	4	13	Y2
$\overline{G2B}$	5	12	Y3
G1	6	11	Y4
Y7	7	10	Y5
GND	8	9	Y6

SN54LS138, SN54S138 . . . FK PACKAGE
(TOP VIEW)

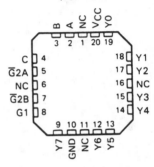

NC — No internal connection

logic symbols†

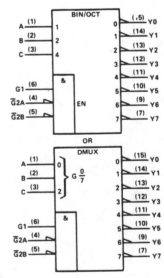

†These symbols are in accordance with ANSI/IEEE Std 91-1984 and IEC Publication 617-12.
Pin numbers shown are for D, J, N, and W packages.

Copyright © 1972, Texas Instruments Incorporated

TEXAS INSTRUMENTS
POST OFFICE BOX 655012 • DALLAS, TEXAS 75265

2

TTL Devices

SN54LS138, SN54S138, SN74LS138, SN74S138A
3-LINE-TO 8-LINE DECODERS/DEMULTIPLEXERS

logic diagram and function table

'LS138, SN54S138, SN74S138A

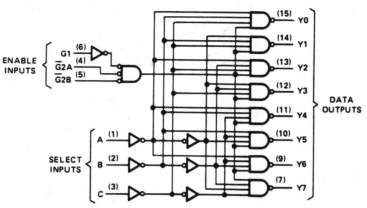

Pin numbers shown are for D, J, N, and W packages.

'LS138, SN54138, SN74S138A
FUNCTION TABLE

INPUTS					OUTPUTS							
ENABLE		SELECT										
G1	G̅2*	C	B	A	Y0	Y1	Y2	Y3	Y4	Y5	Y6	Y7
X	H	X	X	X	H	H	H	H	H	H	H	H
L	X	X	X	X	H	H	H	H	H	H	H	H
H	L	L	L	L	L	H	H	H	H	H	H	H
H	L	L	L	H	H	L	H	H	H	H	H	H
H	L	L	H	L	H	H	L	H	H	H	H	H
H	L	L	H	H	H	H	H	L	H	H	H	H
H	L	H	L	L	H	H	H	H	L	H	H	H
H	L	H	L	H	H	H	H	H	H	L	H	H
H	L	H	H	L	H	H	H	H	H	H	L	H
H	L	H	H	H	H	H	H	H	H	H	H	L

* $\overline{G2} = \overline{G2A} + \overline{G2B}$
H = high level, L = low level, X = irrelevant

TEXAS
INSTRUMENTS
POST OFFICE BOX 655012 • DALLAS, TEXAS 75265

SN54147, SN54148, SN54LS147, SN54LS148, SN74147, SN74148 (TIM9907), SN74LS147, SN74LS148
10-LINE TO 4-LINE AND 8-LINE TO 3-LINE PRIORITY ENCODERS

OCTOBER 1976 – REVISED MARCH 1988

'147, 'LS147

- Encodes 10-Line Decimal to 4-Line BCD

- Applications Include:

  Keyboard Encoding
  Range Selection: '148, 'LS148

- Encodes 8 Data Lines to 3-Line Binary (Octal)

- Applications Include:

  N-Bit Encoding
  Code Converters and Generators

TYPE	TYPICAL DATA DELAY	TYPICAL POWER DISSIPATION
'147	10 ns	225 mW
'148	10 ns	190 mW
'LS147	15 ns	60 mW
'LS148	15 ns	60 mW

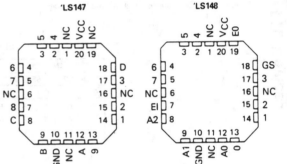

SN54147, SN54LS147,
SN54148, SN54LS148 . . . J OR W PACKAGE
SN74147, SN74148 . . . N PACKAGE
SN74LS147, SN74LS148 . . . D OR N PACKAGE
(TOP VIEW)

SN54LS147, SN54LS148 . . . FK PACKAGE
(TOP VIEW)

NC - No internal connection

description

These TTL encoders feature priority decoding of the inputs to ensure that only the highest-order data line is encoded. The '147 and 'LS147 encode nine data lines to four-line (8-4-2-1) BCD. The implied decimal zero condition requires no input condition as zero is encoded when all nine data lines are at a high logic level. The '148 and 'LS148 encode eight data lines to three-line (4-2-1) binary (octal). Cascading circuitry (enable input EI and enable output EO) has been provided to allow octal expansion without the need for external circuitry. For all types, data inputs and outputs are active at the low logic level. All inputs are buffered to represent one normalized Series 54/74 or 54LS/74LS load, respectively.

'147, 'LS147 FUNCTION TABLE

1	2	3	4	5	6	7	8	9	D	C	B	A
H	H	H	H	H	H	H	H	H	H	H	H	H
X	X	X	X	X	X	X	X	L	L	H	H	L
X	X	X	X	X	X	X	L	H	L	H	H	H
X	X	X	X	X	X	L	H	H	H	L	L	L
X	X	X	X	X	L	H	H	H	H	L	L	H
X	X	X	X	L	H	H	H	H	H	L	H	L
X	X	X	L	H	H	H	H	H	H	L	H	H
X	X	L	H	H	H	H	H	H	H	H	L	L
X	L	H	H	H	H	H	H	H	H	H	L	H
L	H	H	H	H	H	H	H	H	H	H	H	L

(Columns: INPUTS 1–9, OUTPUTS D C B A)

'148, 'LS148 FUNCTION TABLE

EI	0	1	2	3	4	5	6	7	A2	A1	A0	GS	EO
H	X	X	X	X	X	X	X	X	H	H	H	H	H
L	H	H	H	H	H	H	H	H	H	H	H	H	L
L	X	X	X	X	X	X	X	L	L	L	L	L	H
L	X	X	X	X	X	X	L	H	L	L	H	L	H
L	X	X	X	X	X	L	H	H	L	H	L	L	H
L	X	X	X	X	L	H	H	H	L	H	H	L	H
L	X	X	X	L	H	H	H	H	H	L	L	L	H
L	X	X	L	H	H	H	H	H	H	L	H	L	H
L	X	L	H	H	H	H	H	H	H	H	L	L	H
L	L	H	H	H	H	H	H	H	H	H	H	L	H

(Columns: INPUTS EI, 0–7, OUTPUTS A2 A1 A0 GS EO)

H = high logic level, L = low logic level, X = irrelevant

TEXAS
INSTRUMENTS

POST OFFICE BOX 655012 • DALLAS, TEXAS 75265

TTL Devices

2

SN54147, SN54148, SN54LS147, SN54LS148,
SN74147, SN74148 (TIM9907), SN74LS147, SN74LS148
10-LINE TO 4-LINE AND 8-LINE TO 3-LINE PRIORITY ENCODERS

logic symbols[†]

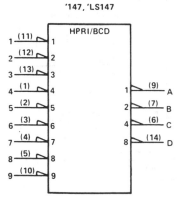

'147, 'LS147

'148, 'LS148

[†]These symbols are in accordance with ANSI/IEEE Std. 91-1984 and IEC Publication 617-12.

Pin numbers shown are for D, J, N, and W packages.

logic diagrams

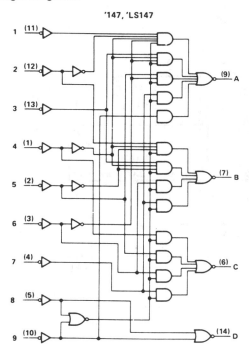

'147, 'LS147

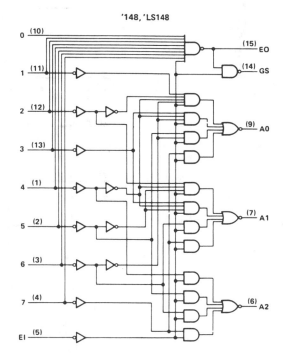

'148, 'LS148

Pin numbers shown are for D, J, N, and W packages.

TEXAS
INSTRUMENTS

POST OFFICE BOX 655012 • DALLAS, TEXAS 75265

SN54147, SN54148 (TIM9907), SN54LS147, SN54LS148,
SN74147, SN74148, SN74LS147, SN74LS148
10-LINE TO 4-LINE AND 8-LINE TO 3-LINE PRIORITY ENCODERS

TYPICAL APPLICATION DATA

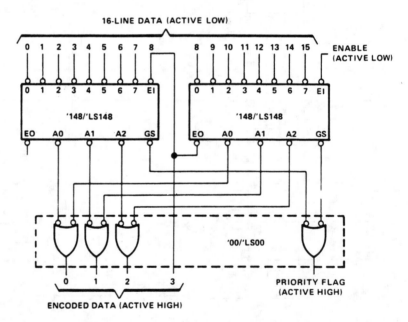

Since the '147/'LS147 and '148/'LS148 are combinational logic circuits, wrong addresses can appear during input transients. Moreover, for the '148/'LS148 a change from high to low at input EI can cause a transient low on the GS output when all inputs are high. This must be considered when strobing the outputs.

TEXAS
INSTRUMENTS

POST OFFICE BOX 655012 • DALLAS, TEXAS 75265

SN54150, SN54151A, SN54LS151, SN54S151, SN74150, SN74151A, SN74LS151, SN74S151
DATA SELECTORS/MULTIPLEXERS

DECEMBER 1972 – REVISED MARCH 1988

- '150 Selects One-of-Sixteen Data Sources
- Others Select One-of-Eight Data Sources
- All Perform Parallel-to-Serial Conversion
- All Permit Multiplexing from N Lines to One Line
- Also For Use as Boolean Function Generator
- Input-Clamping Diodes Simplify System Design
- Fully Compatible with Most TTL Circuits

TYPE	TYPICAL AVERAGE PROPAGATION DELAY TIME DATA INPUT TO W OUTPUT	TYPICAL POWER DISSIPATION
'150	13 ns	200 mW
'151A	8 ns	145 mW
'LS151	13 ns	30 mW
'S151	4.5 ns	225 mW

description

These monolithic data selectors/multiplexers contain full on-chip binary decoding to select the desired data source. The '150 selects one-of-sixteen data sources; the '151A, 'LS151, and 'S151 select one-of-eight data sources. The '150, '151A, 'LS151, and 'S151 have a strobe input which must be at a low logic level to enable these devices. A high level at the strobe forces the W output high, and the Y output (as applicable) low.

The '150 has only an inverted W output; the '151A, 'LS151, and 'S151 feature complementary W and Y outputs.

The '151A and '152A incorporate address buffers that have symmetrical propagation delay times through the complementary paths. This reduces the possibility of transients occurring at the output(s) due to changes made at the select inputs, even when the '151A outputs are enabled (i.e., strobe low).

SN54150 . . . J OR W PACKAGE
SN74150 . . . N PACKAGE
(TOP VIEW)

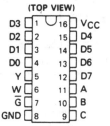

SN54151A, SN54LS151, SN54S151 . . . J OR W PACKAGE
SN74151A . . . N PACKAGE
SN74LS151, SN74S151 . . . D OR N PACKAGE
(TOP VIEW)

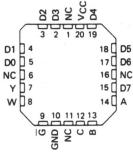

SN54LS151, SN54S151 . . . FK PACKAGE
(TOP VIEW)

NC - No internal connection

2

TTL Devices

TEXAS
INSTRUMENTS
POST OFFICE BOX 655012 • DALLAS, TEXAS 75265

SN54150, SN54151A, SN54LS151, SN54S151, SN74150, SN74151A, SN74LS151, SN74S151
DATA SELECTORS/MULTIPLEXERS

logic symbols†

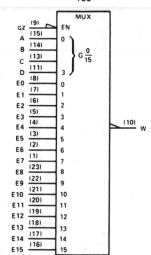

'150

'151A, 'LS151, 'S151

†These symbols are in accordance with ANSI/IEEE Std. 91-1984 and IEC Publication 617-12.
Pin numbers shown are D, J, N, and W packages.

'150

FUNCTION TABLE

INPUTS					OUTPUT
SELECT				STROBE	W
D	C	B	A	$\overline{G}$	
X	X	X	X	H	H
L	L	L	L	L	$\overline{E0}$
L	L	L	H	L	$\overline{E1}$
L	L	H	L	L	$\overline{E2}$
L	L	H	H	L	$\overline{E3}$
L	H	L	L	L	$\overline{E4}$
L	H	L	H	L	$\overline{E5}$
L	H	H	L	L	$\overline{E6}$
L	H	H	H	L	$\overline{E7}$
H	L	L	L	L	$\overline{E8}$
H	L	L	H	L	$\overline{E9}$
H	L	H	L	L	$\overline{E10}$
H	L	H	H	L	$\overline{E11}$
H	H	L	L	L	$\overline{E12}$
H	H	L	H	L	$\overline{E13}$
H	H	H	L	L	$\overline{E14}$
H	H	H	H	L	$\overline{E15}$

'151A, 'LS151, 'S151

FUNCTION TABLE

INPUTS				OUTPUTS	
SELECT			STROBE	Y	W
C	B	A	$\overline{G}$		
X	X	X	H	L	H
L	L	L	L	D0	$\overline{D0}$
L	L	H	L	D1	$\overline{D1}$
L	H	L	L	D2	$\overline{D2}$
L	H	H	L	D3	$\overline{D3}$
H	L	L	L	D4	$\overline{D4}$
H	L	H	L	D5	$\overline{D5}$
H	H	L	L	D6	$\overline{D6}$
H	H	H	L	D7	$\overline{D7}$

H = high level, L = low level, X = irrelevant
$\overline{E0}$, $\overline{E1}$ . . . $\overline{E15}$ = the complement of the level of the respective E input
D0, D1 . . . D7 = the level of the D respective input

TEXAS INSTRUMENTS
POST OFFICE BOX 655012 • DALLAS, TEXAS 75265

SN54150, SN54151A, SN54LS151, SN54S151, SN74150, SN74151A, SN74LS151, SN74S151
DATA SELECTORS/MULTIPLEXERS

logic diagrams (positive logic)

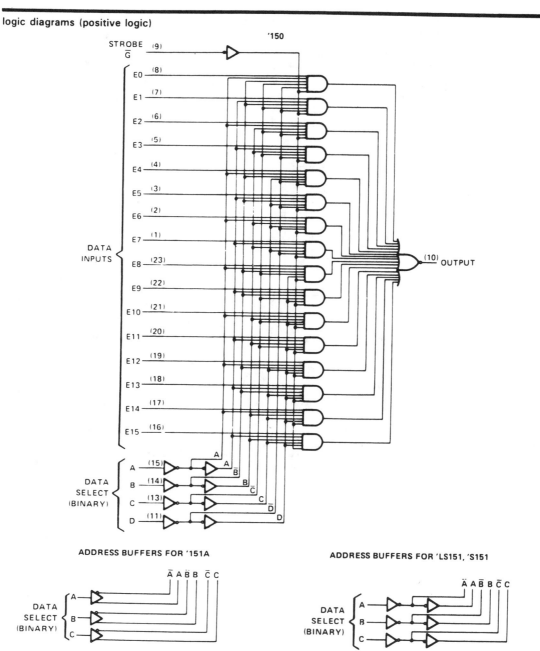

ADDRESS BUFFERS FOR '151A

ADDRESS BUFFERS FOR 'LS151, 'S151

Pin numbers shown are for D, J, N, and W packages.

TEXAS
INSTRUMENTS
POST OFFICE BOX 655012 • DALLAS, TEXAS 75265

'151A, 'LS151, 'S151

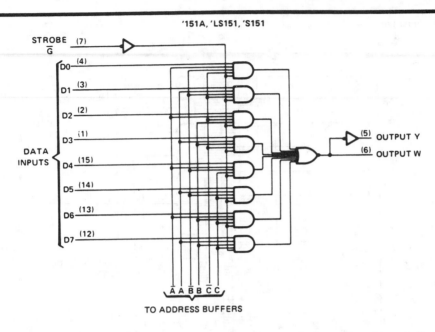

TO ADDRESS BUFFERS

2

TTL Devices

absolute maximum ratings over operating free-air temperature range (unless otherwise noted)

Supply voltage, V_{CC} (see Note 1) . . . . 7 V
Input voltage (see Note 2): '150, '151A, 'S151 . . . . 5.5 V
                            'LS151 . . . . 7 V
Operating free-air temperature range: SN54' . . . . −55°C to 125°C
                                        SN74' . . . . 0°C to 70°C
Storage temperature range . . . . −65°C to 150°C

NOTES: 1: Voltage values are with respect to network ground terminal.
         2. For the '150, input voltages must be zero or positive with respect to network ground terminal.

TEXAS
INSTRUMENTS
POST OFFICE BOX 655012 • DALLAS, TEXAS 75265

- Permits Multiplexing from N lines to 1 line
- Performs Parallel-to-Serial Conversion
- Strobe (Enable) Line Provided for Cascading (N lines to n lines)
- High-Fan-Out, Low-Impedance, Totem-Pole Outputs
- Fully Compatible with most TTL Circuits

TYPE	TYPICAL AVERAGE PROPAGATION DELAY TIMES			TYPICAL POWER DISSIPATION
	FROM DATA	FROM STROBE	FROM SELECT	
'153	14 ns	17 ns	22 ns	180 mW
'LS153	14 ns	19 ns	22 ns	31 mW
'S153	6 ns	9.5 ns	12 ns	225 mW

description

Each of these monolithic, data selectors/multiplexers contains inverters and drivers to supply fully complementary, on-chip, binary decoding data selection to the AND-OR gates. Separate strobe inputs are provided for each of the two four-line sections.

FUNCTION TABLE

SELECT INPUTS		DATA INPUTS				STROBE	OUTPUT
B	A	C0	C1	C2	C3	G̅	Y
X	X	X	X	X	X	H	L
L	L	L	X	X	X	L	L
L	L	H	X	X	X	L	H
L	H	X	L	X	X	L	L
L	H	X	H	X	X	L	H
H	L	X	X	L	X	L	L
H	L	X	X	H	X	L	H
H	H	X	X	X	L	L	L
H	H	X	X	X	H	L	H

Select inputs A and B are common to both sections.
H = high level, L = low level, X = irrelevant

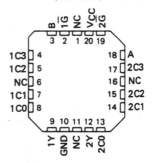

SN54153, SN54LS153, SN54S153 . . . J OR W PACKAGE
SN74153 . . . N PACKAGE
SN74LS153, SN74S153 . . . D OR N PACKAGE
(TOP VIEW)

1G̅	1	16	V_{CC}
B	2	15	2G̅
1C3	3	14	A
1C2	4	13	2C3
1C1	5	12	2C2
1C0	6	11	2C1
1Y	7	10	2C0
GND	8	9	2Y

SN54LS153, SN54S153 . . . FK PACKAGE
(TOP VIEW)

NC - No internal connection

2

TTL Devices

absolute maximum ratings over operating free-air temperature range (unless otherwise noted)

Supply voltage, V_{CC} (See Note 1) . . . . 7 V
Input voltage: '153, 'S153 . . . . 5.5 V
              'LS153 . . . . 7 V
Operating free-air temperature range: SN54' . . . . −55°C to 125°C
                           SN74' . . . . 0°C to 70°C
Storage temperature range . . . . −65°C to 150°C

NOTE 1: Voltage values are with respect to network ground terminal.

TEXAS
INSTRUMENTS

POST OFFICE BOX 655012 • DALLAS, TEXAS 75265

SN54153, SN54LS153, SN54S153
SN74153, SN74LS153, SN74S153
DUAL 4-LINE TO 1-LINE DATA SELECTORS/MULTIPLEXERS

logic symbol†

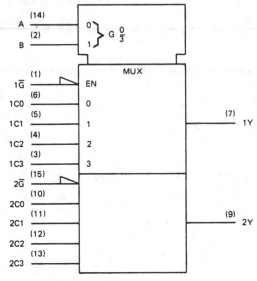

†This symbol is in accordance with ANSI/IEEE Std. 91-1984 and IEC
  Publication 617-12.

logic diagrams (positive logic)

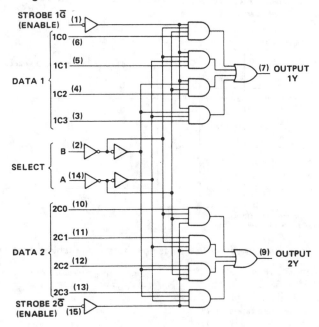

Pin numbers shown are for D, J, N, and W packages.

TEXAS
INSTRUMENTS
POST OFFICE BOX 655012 • DALLAS. TEXAS 75265

'160,'161,'LS160A,'LS161A . . . SYNCHRONOUS COUNTERS WITH DIRECT CLEAR
'162,'163,'LS162A,'LS163A,'S162,'S163 . . . FULLY SYNCHRONOUS COUNTERS

- Internal Look-Ahead for Fast Counting
- Carry Output for n-Bit Cascading
- Synchronous Counting
- Synchronously Programmable
- Load Control Line
- Diode-Clamped Inputs

TYPE	TYPICAL PROPAGATION TIME, CLOCK TO Q OUTPUT	TYPICAL MAXIMUM CLOCK FREQUENCY	TYPICAL POWER DISSIPATION
'160 thru '163	14 ns	32 MHz	305 mW
'LS162A thru 'LS163A	14 ns	32 MHz	93 mW
'S162 and 'S163	9 ns	70 MHz	475 mW

SERIES 54', 54LS' 54S' . . . J OR W PACKAGE
SERIES 74' . . . N PACKAGE
SERIES 74LS', 74S' . . . D OR N PACKAGE
(TOP VIEW)

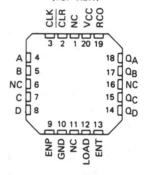

$\overline{CLR}$	1	16	V_{CC}
CLK	2	15	RCO
A	3	14	Q_A
B	4	13	Q_B
C	5	12	Q_C
D	6	11	Q_D
ENP	7	10	ENT
GND	8	9	LOAD

NC—No internal connection

SERIES 54LS', 54S' . . . FK PACKAGE
(TOP VIEW)

NC—No internal connection

description

These synchronous, presettable counters feature an internal carry look-ahead for application in high-speed counting designs. The '160, '162, 'LS160A, 'LS162A, and 'S162 are decade counters and the '161, '163, 'LS161A, 'LS163A, and 'S163 are 4-bit binary counters. Synchronous operation is provided by having all flip-flops clocked simultaneously so that the outputs change coincident with each other when so instructed by the count-enable inputs and internal gating. This mode of operation eliminates the output counting spikes that are normally associated with asynchronous (ripple clock) counters, however counting spikes may occur on the (RCO) ripple carry output. A buffered clock input triggers the four flip-flops on the rising edge of the clock input waveform.

These counters are fully programmable; that is, the outputs may be preset to either level. As presetting is synchronous, setting up a low level at the load input disables the counter and causes the outputs to agree with the setup data after the next clock pulse regardless of the levels of the enable inputs. Low-to-high transitions at the load input of the '160 thru '163 should be avoided when the clock is low if the enable inputs are high at or before the transition. This restriction is not applicable to the 'LS160A thru 'LS163A or 'S162 or 'S163. The clear function for the '160, '161, 'LS160A, and 'LS161A is asynchronous and a low level at the clear input sets all four of the flip-flop outputs low regardless of the levels of clock, load, or enable inputs. The clear function for the '162, '163, 'LS162A, 'LS163A, 'S162, and 'S163 is synchronous and a low level at the clear input sets all four of the flip-flop outputs low after the next clock pulse, regardless of the levels of the enable inputs. This synchronous clear allows the count length to be modified easily as decoding the maximum count desired can be accomplished with one external NAND gate. The gate output is connected to the clear input to synchronously clear the counter to 0000 (LLLL). Low-to-high transitions at the clear input of the '162 and '163 should be avoided when the clock is low if the enable and load inputs are high at or before the transition.

TEXAS
INSTRUMENTS

POST OFFICE BOX 655012 • DALLAS, TEXAS 75265

TTL Devices

2

SN54160 THRU SN54163, SN54LS160A THRU SN54LS163A, SN54S162, SN54S163, SN74160 THRU SN74163, SN74LS160A THRU SN74LS163A, SN74S162, SN74S163 SYNCHRONOUS 4-BIT COUNTERS

The carry look-ahead circuitry provides for cascading counters for n-bit synchronous applications without additional gating. Instrumental in accomplishing this function are two count-enable inputs and a ripple carry output. Both count-enable inputs (P and T) must be high to count, and input T is fed forward to enable the ripple carry output. The ripple carry output thus enabled will produce a high-level output pulse with a duration approximately equal to the high-level portion of the Q_A output. This high-level overflow ripple carry pulse can be used to enable successive cascaded stages. High-to-low-level transitions at the enable P or T inputs of the '160 thru '163 should occur only when the clock input is high. Transitions at the enable P or T inputs of the 'LS160A thru 'LS163A or 'S162 and 'S163 are allowed regardless of the level of the clock input.

'LS160A thru 'LS163A, 'S162 and 'S163 feature a fully independent clock circuit. Changes at control inputs (enable P or T, or load) that will modify the operating mode have no effect until clocking occurs. The function of the counter (whether enabled, disabled, loading, or counting) will be dictated solely by the conditions meeting the stable setup and hold times.

logic symbols†

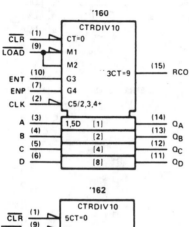

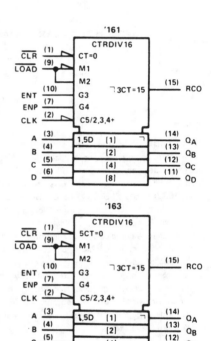

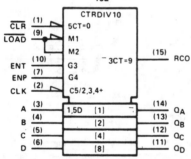

†These symbols are in accordance with ANSI/IEEE Std 91-1984 and IEC Publication 617-12.

Pin numbers shown are for D, J, N, and W packages.

logic symbols (continued) [†]

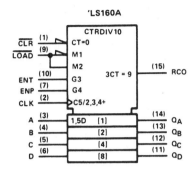

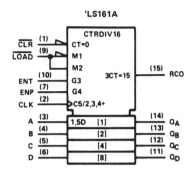

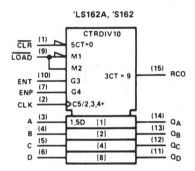

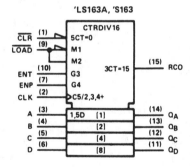

[†]These symbols are in accordance with ANSI/IEEE Std 91-1984 and
IEC Publication 617-12.

Pin numbers shown are for D, J, N, and W packages.

TEXAS
INSTRUMENTS
POST OFFICE BOX 655012 • DALLAS, TEXAS 75265

2

TTL Devices

logic diagram (positive logic)

SN54LS160A, SN74LS160A SYNCHRONOUS
DECADE COUNTERS

SN54LS162A, SN74LS162A synchronous decade counters are similar; however the clear is synchronous as shown for the SN54LS163A, SN74LS163A binary counters at right.

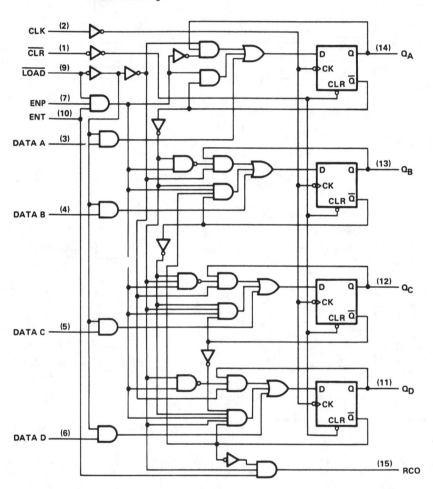

Pin numbers shown are for D, J, N, and W packages.

TEXAS
INSTRUMENTS

POST OFFICE BOX 655012 • DALLAS, TEXAS 75265

logic diagram (positive logic)

SN54LS163A, SN74LS163A SYNCHRONOUS
BINARY COUNTERS

SN54LS161A, SN74LS161A synchronous binary counters are similar; however, the clear is asynchronous as shown for the SN54LS160A, SN74LS160A decade counters at left.

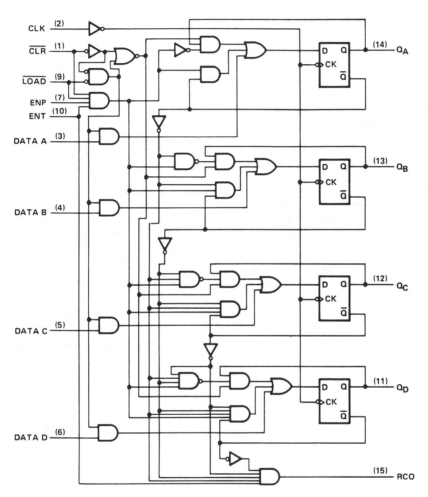

Pin numbers shown are for D, J, N, and W packages.

TEXAS
INSTRUMENTS
POST OFFICE BOX 655012 • DALLAS, TEXAS 75265

'160, '162, 'LS160A, 'LS162A, 'S162 DECADE COUNTERS

typical clear, preset, count, and inhibit sequences

Illustrated below is the following sequence:

1. Clear outputs to zero ('160 and 'LS160A are asynchronous; '162, 'LS162A, and 'S162 are synchronous)
2. Preset to BCD seven
3. Count to eight, nine, zero, one, two, and three
4. Inhibit

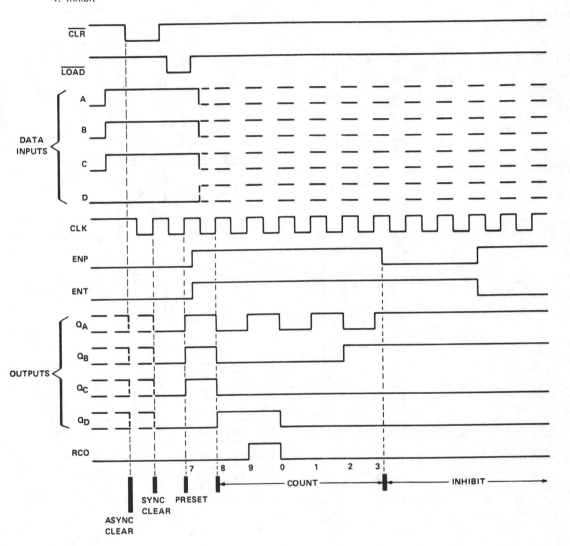

TEXAS
INSTRUMENTS

POST OFFICE BOX 655012 • DALLAS, TEXAS 75265

'161, 'LS161A, '163, 'LS163A, 'S163 BINARY COUNTERS

typical clear, preset, count, and inhibit sequences

Illustrated below is the following sequence:
1. Clear outputs to zero ('161 and 'LS161A are asynchronous; '163, 'LS163A, and 'S163 are synchronous)
2. Preset to binary twelve
3. Count to thirteen, fourteen fifteen, zero, one, and two
4. Inhibit

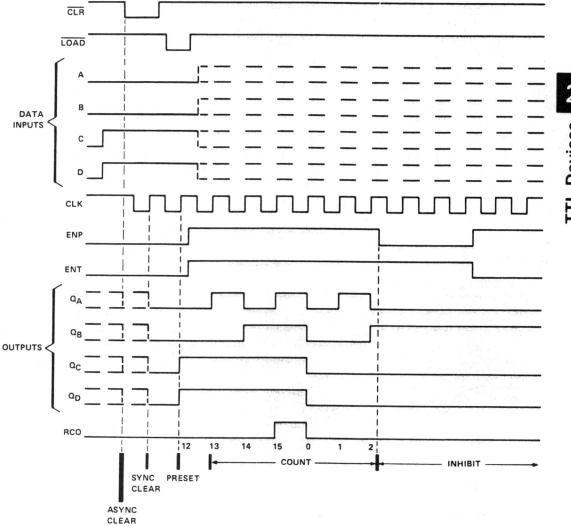

schematics of inputs and outputs

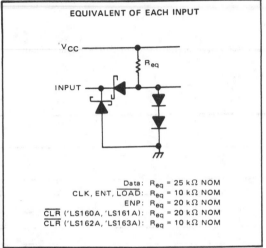

EQUIVALENT OF EACH INPUT

Data: R_{eq} = 25 kΩ NOM
CLK, ENT, $\overline{\text{LOAD}}$: R_{eq} = 10 kΩ NOM
ENP: R_{eq} = 20 kΩ NOM
$\overline{\text{CLR}}$ ('LS160A, 'LS161A): R_{eq} = 20 kΩ NOM
$\overline{\text{CLR}}$ ('LS162A, 'LS163A): R_{eq} = 10 kΩ NOM

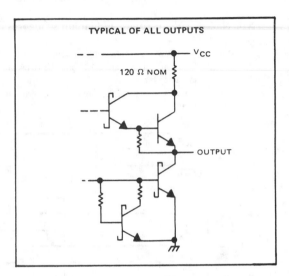

TYPICAL OF ALL OUTPUTS

absolute maximum ratings over operating free-air temperature range (unless otherwise noted)

Supply voltage, V_{CC} (see Note 7) . . . . 7 V
Input voltage . . . . 7 V
Operating free-air temperature range: SN54LS' Circuits . . . . −55°C to 125°C
SN74LS' Circuits . . . . 0°C to 70°C
Storage temperature range . . . . −65°C to 150°C

NOTE 7: Voltage values are with respect to network ground terminal.

recommended operating conditions

			SN54LS'			SN74LS'			UNIT
			MIN	NOM	MAX	MIN	NOM	MAX	
V_{CC}	Supply voltage		4.5	5	5.5	4.75	5	5.25	V
I_{OH}	High-level output current				−400			−400	µA
I_{OL}	Low-level output current				4			8	mA
f_{clock}	Clock frequency		0		25	0		25	MHz
$t_{w(clock)}$	Width of clock pulse		25			25			ns
$t_{w(clear)}$	Width of clear pulse		20			20			ns
t_{su}	Setup time, (see Figures 1 and 2)	Data inputs A, B, C, D	20			20			ns
		ENP or ENT	20			20			
		$\overline{\text{LOAD}}$	20			20			
		$\overline{\text{LOAD}}$ inactive state	20			20			
		$\overline{\text{CLR}}$ †	20			20			
		$\overline{\text{CLR}}$ inactive state	25			25			
t_h	Hold time at any input		3			3			ns
T_A	Operating free-air temperature		−55		125	0		70	°C

† This applies only for 'LS162 and 'LS163, which have synchronous clear inputs.

TEXAS
INSTRUMENTS
POST OFFICE BOX 655012 • DALLAS, TEXAS 75265

electrical characteristics over recommended operating free-air temperature range (unless otherwise noted)

	PARAMETER		TEST CONDITIONS[†]	SN54LS' MIN	TYP[‡]	MAX	SN74LS' MIN	TYP[‡]	MAX	UNIT
V_{IH}	High-level input voltage			2			2			V
V_{IL}	Low-level input voltage					0.7			0.8	V
V_{IK}	Input clamp voltage		V_{CC} = MIN, I_I = −18 mA			−1.5			−1.5	V
V_{OH}	High-level output voltage		V_{CC} = MIN, V_{IH} = 2 V, V_{IL} = V_{IL} max, I_{OH} = −400 µA	2.5	3.4		2.7	3.4		V
V_{OL}	Low-level output voltage		V_{CC} = MIN, V_{IH} = 2 V, V_{IL} = V_{IL} max I_{OL} = 4 mA		0.25	0.4		0.25	0.4	V
			I_{OL} = 8 mA					0.35	0.5	
I_I	Input current at maximum input voltage	Data or ENP	V_{CC} = MAX, V_I = 7 V			0.1			0.1	mA
		$\overline{LOAD}$, CLK, or ENT				0.2			0.2	
		$\overline{CLR}$ ('LS160A, 'LS161A)				0.1			0.1	
		$\overline{CLR}$ ('LS162A, 'LS163A)				0.2			0.2	
I_{IH}	High-level input current	Data or ENP	V_{CC} = MAX, V_I = 2.7 V			20			20	µA
		$\overline{LOAD}$, CLK, or ENT				40			40	
		$\overline{CLR}$ ('LS160A, 'LS161A)				20			20	
		$\overline{CLR}$ ('LS162A, 'LS163A)				40			40	
I_{IL}	Low-level input current	Data or ENP	V_{CC} = MAX, V_I = 0.4 V			−0.4			−0.4	mA
		$\overline{LOAD}$, CLK, or ENT				−0.8			−0.8	
		$\overline{CLR}$ ('LS160A, 'LS161A)				−0.4			−0.4	
		$\overline{CLR}$ ('LS162A, 'LS163A)				−0.8			−0.8	
I_{OS}	Short-circuit output current[§]		V_{CC} = MAX	−20		−100	−20		−100	mA
I_{CCH}	Supply current, all outputs high		V_{CC} = MAX, See Note 3		18	31		18	31	mA
I_{CCL}	Supply current, all outputs low		V_{CC} = MAX, See Note 4		19	32		19	32	mA

[†]For conditions shown as MIN or MAX, use the appropriate value specified under recommended operating conditions.
[‡]All typical values are at V_{CC} = 5 V, T_A = 25°C.
[§]Not more than one output should be shorted at a time, and duration of the short-circuit should not exceed one second.
NOTES: 3. I_{CCH} is measured with the load input high, then again with the load input low, with all other inputs high and all outputs open.
4. I_{CCL} is measured with the clock input high, then again with the clock input low, with all other inputs low and all outputs open.

switching characteristics, V_{CC} = 5 V, T_A = 25°C

PARAMETER[¶]	FROM (INPUT)	TO (OUTPUT)	TEST CONDITIONS	MIN	TYP	MAX	UNIT
f_{max}				25	32		MHz
t_{PLH}	CLK	RCO			20	35	ns
t_{PHL}					18	35	
t_{PLH}	CLK	Any	C_L = 15 pF,		13	24	ns
t_{PHL}	($\overline{LOAD}$ input high)	Q	R_L = 2 kΩ,		18	27	
t_{PLH}	CLK	Any	See figures		13	24	ns
t_{PHL}	($\overline{LOAD}$ input low)	Q	1 and 2 and		18	27	
t_{PLH}	ENT	RCO	Note 8		9	14	ns
t_{PHL}					9	14	
t_{PHL}	$\overline{CLR}$	Any Q			20	28	ns

[¶]f_{max} = Maximum clock frequency
t_{PLH} = propagation delay time, low-to-high-level output.
t_{PHL} = propagation delay time, high-to-low-level output.
NOTE 8: Propagation delay for clearing is measured from the clear input for the 'LS160A and 'LS161A or from the clock transition for the 'LS162A and 'LS163A.

TEXAS
INSTRUMENTS
POST OFFICE BOX 655012 • DALLAS, TEXAS 75265

2-507

2

TTL Devices

PARAMETER MEASUREMENT INFORMATION

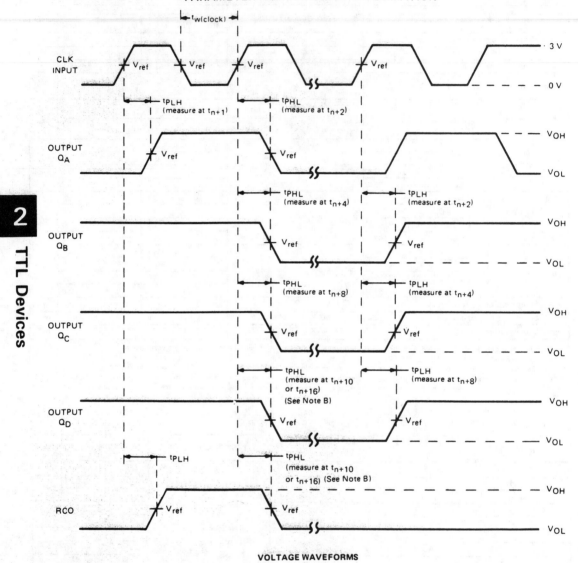

VOLTAGE WAVEFORMS

NOTES: A. The input pulses are supplied by a generator having the following characteristics: PRR ≤ 1 MHz, duty cycle ≤ 50%, $Z_{out} \approx 50\ \Omega$; for '160 thru '163, t_r ≤ 10 ns, t_f ≤ 10 ns; for 'LS160A thru 'LS163A t_r ≤ 15 ns, t_f ≤ 6 ns; and for 'S162, 'S163, t_r ≤ 2.5 ns, t_f ≤ 2.5 ns. Vary PRR to measure f_{max}.
  B. Outputs Q_D and carry are tested at t_{n+10} for '160, '162, 'LS160A, 'LS162A, and 'S162, and at t_{n+16} for '161, '163, 'LS161A, 'LS163A, and 'S163, where t_n is the bit time when all outputs are low.
  C. For '160 thru '163, 'S162, and 'S163, V_{ref} = 1.5 V; for 'LS160A thru 'LS163A, V_{ref} = 1.3 V.

FIGURE 1—SWITCHING TIMES

TEXAS
INSTRUMENTS

POST OFFICE BOX 655012 • DALLAS, TEXAS 75265

PARAMETER MEASUREMENT INFORMATION

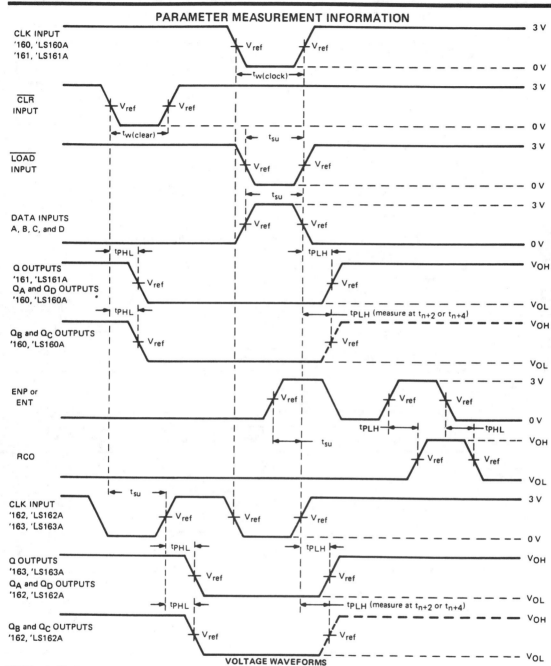

2

TTL Devices

VOLTAGE WAVEFORMS

NOTES: A. The input pulses are supplied by generators having the following characteristics: PRR ≤ 1 MHz, duty cycle ≤ 50%, $Z_{out} \approx 50\ \Omega$; for '160 thru '163, $t_r \leq 10$ ns, $t_f \leq 10$ ns; and for 'LS160A thru 'LS163A, $t_r \leq 15$ ns, $t_f \leq 6$ ns.
   B. Enable P and enable T setup times are measured at t_{n+0}.
   C. For '160 thru '163, $V_{ref} = 1.5$ V; for 'LS160A thru 'LS163A, $V_{ref} = 1.3$ V.

FIGURE 2—SWITCHING TIMES

TEXAS
INSTRUMENTS
POST OFFICE BOX 655012 • DALLAS, TEXAS 75265

2

TTL Devices

PARAMETER MEASUREMENT INFORMATION

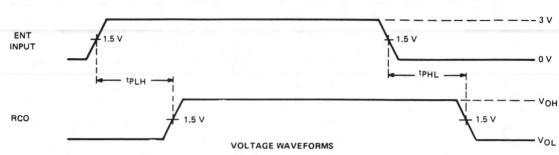

VOLTAGE WAVEFORMS

NOTES: A. The input pulse is supplied by a generator having the following characteristics: $t_r \leq 2.5$ ns, $t_f \leq 2.5$ ns, PRR 1 MHz, duty cycle 50%, $Z_{out} \approx 50\ \Omega$.

B. t_{PLH} and t_{PHL} from enable T input to carry output assume that the counter is at the maximum count (Q_A and Q_D high for 'S162, all Q outputs high for 'S163).

FIGURE 3—PROPAGATION DELAY TIMES FROM ENABLE T INPUT TO CARRY OUTPUT

VOLTAGE WAVEFORMS

NOTE A: The input pulses are supplied by generators having the following characteristics: t_r 2.5 ns, t_f 2.5 ns, PRR 1 MHz, duty cycle 50%, $Z_{out} \approx 50\ \Omega$.

FIGURE 4—PULSE WIDTHS, SETUP TIMES, HOLD TIMES, AND RELEASE TIME

TEXAS INSTRUMENTS

POST OFFICE BOX 655012 • DALLAS, TEXAS 75265

TYPICAL APPLICATION DATA

This application demonstrates how the ripple mode carry circuit (Figure 1) and the carry-look-ahead circuit (Figure 2) can be used to implement a high-speed N-bit counter. The '160, '162, 'LS160A, 'LS162A, or 'S162 will count in BCD and the '161, '163, 'LS161A, 'LS163A, or 'S163 will count in binary. When additional stages are added the f_{MAX} decreases in Figure 1, but remains unchanged in Figure 2.

N-BIT SYNCHRONOUS COUNTERS

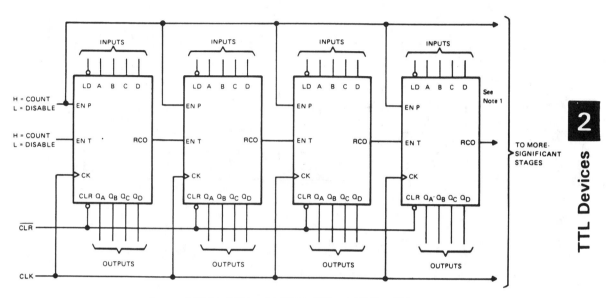

$$f_{MAX} = 1/(\text{CLK to RCO } t_{PLH}) + (\text{ENT to RCO } t_{PLH})(N-2) + (\text{ENT } t_{su})$$

FIGURE 1

TEXAS
INSTRUMENTS
POST OFFICE BOX 655012 • DALLAS, TEXAS 75265

TYPICAL APPLICATION DATA

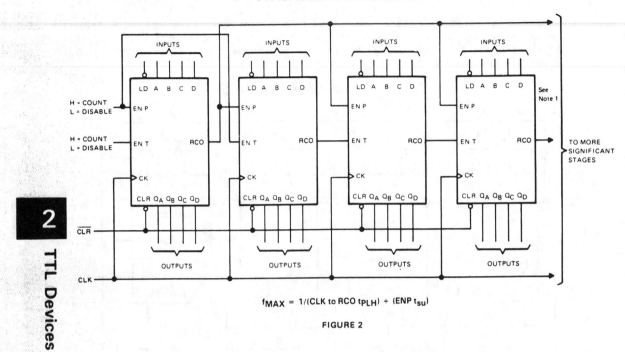

$$f_{MAX} = 1/(\text{CLK to RCO } t_{PLH}) + (\text{ENP } t_{su})$$

FIGURE 2

TEXAS
INSTRUMENTS

POST OFFICE BOX 655012 • DALLAS, TEXAS 75265

- **Gated Serial Inputs**
- **Fully Buffered Clock and Serial Inputs**
- **Asynchronous Clear**

TYPE	TYPICAL MAXIMUM CLOCK FREQUENCY	TYPICAL POWER DISSIPATION
'164	36 MHz	21 mW per bit
'LS164	36 MHz	10 mW per bit

description

These 8-bit shift registers feature gated serial inputs and an asynchronous clear. The gated serial inputs (A and B) permit complete control over incoming data as a low at either input inhibits entry of the new data and resets the first flip-flop to the low level at the next clock pulse. A high-level input enables the other input which will then determine the state of the first flip-flop. Data at the serial inputs may be changed while the clock is high or low, but only information meeting the setup-time requirements will be entered. Clocking occurs on the low-to-high-level transition of the clock input. All inputs are diode-clamped to minimize transmission-line effects.

The SN54164 and SN54LS164 are characterized for operation over the full military temperature range of −55°C to 125°C. The SN74164 and SN74LS164 are characterized for operation from 0°C to 70°C.

SN54164, SN54LS164 . . . J OR W PACKAGE
SN74164 . . . N PACKAGE
SN74LS164 . . . D OR N PACKAGE
(TOP VIEW)

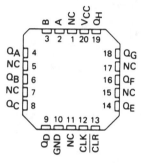

```
         ___  ___
A   [ 1       14 ] V_CC
B   [ 2       13 ] Q_H
Q_A [ 3       12 ] Q_G
Q_B [ 4       11 ] Q_F
Q_C [ 5       10 ] Q_E
Q_D [ 6        9 ] CLR
GND [ 7        8 ] CLK
```

SN54LS164 . . . FK PACKAGE
(TOP VIEW)

```
          B   A  NC V_CC Q_H
          3   2   1  20  19
    Q_A [ 4              18 ] Q_G
    NC  [ 5              17 ] NC
    Q_B [ 6              16 ] Q_F
    NC  [ 7              15 ] NC
    Q_C [ 8              14 ] Q_E
          9  10  11  12  13
         Q_D GND NC CLK CLR
```

NC — No internal connection

FUNCTION TABLE

INPUTS				OUTPUTS			
$\overline{CLEAR}$	CLOCK	A	B	Q_A	Q_B	...	Q_H
L	X	X	X	L	L		L
H	L	X	X	Q_{A0}	Q_{B0}		Q_{H0}
H	↑	H	H	H	Q_{An}		Q_{Gn}
H	↑	L	X	L	Q_{An}		Q_{Gn}
H	↑	X	L	L	Q_{An}		Q_{Gn}

H = high level (steady state), L = low level (steady state)
X = irrelevant (any input, including transitions)
↑ = transition from low to high level.
Q_{A0}, Q_{B0}, Q_{H0} = the level of Q_A, Q_B, or Q_H, respectively, before the indicated steady-state input conditions were established.
Q_{An}, Q_{Gn} = the level of Q_A or Q_G before the most-recent ↑ transition of the clock; indicates a one-bit shift.

schematics of inputs and outputs

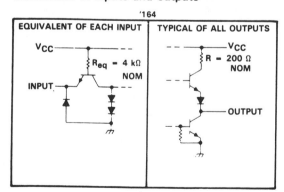

'164

EQUIVALENT OF EACH INPUT	TYPICAL OF ALL OUTPUTS
V_CC R_eq = 4 kΩ NOM INPUT	V_CC R = 200 Ω NOM OUTPUT

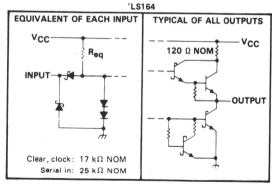

'LS164

EQUIVALENT OF EACH INPUT	TYPICAL OF ALL OUTPUTS
V_CC R_eq INPUT Clear, clock: 17 kΩ NOM Serial in: 25 kΩ NOM	V_CC 120 Ω NOM OUTPUT

2 / TTL Devices

typical clear, shift, and clear sequences

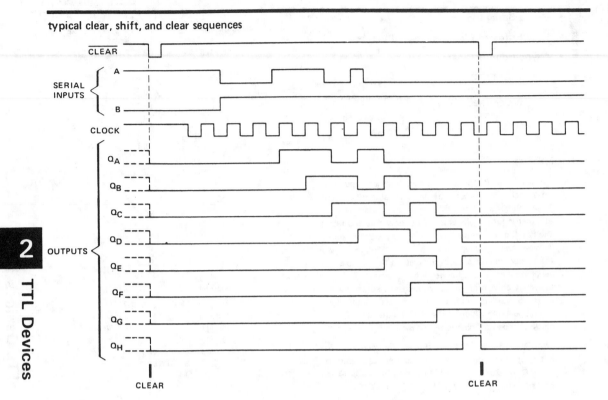

logic symbol[†]

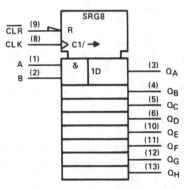

[†]This symbol is in accordance with ANSI/IEEE Std. 91-1984 and IEC Publication 617-12.
Pin numbers shown are for D, J, N, and W packages.

TEXAS
INSTRUMENTS
POST OFFICE BOX 655012 • DALLAS, TEXAS 75265

2

TTL Devices

logic diagram (positive logic)

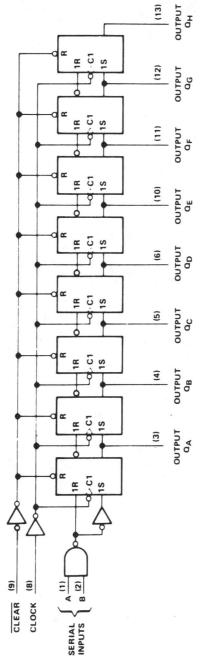

Pin numbers shown are for D, J, N, and W packages.

TEXAS
INSTRUMENTS
POST OFFICE BOX 655012 • DALLAS, TEXAS 75265

- **Synchronous Load**
- **Direct Overriding Clear**
- **Parallel to Serial Conversion**

TYPE	TYPICAL MAXIMUM CLOCK FREQUENCY	TYPICAL POWER DISSIPATION
'166	35 MHz	360 mW
'LS166A	35 MHz	100 mW

description

The '166 and 'LS166A 8-bit shift registers are compatible with most other TTL logic families. All '166 and 'LS166A inputs are buffered to lower the drive requirements to one Series 54/74 or Series 54LS/74LS standard load, respectively. Input clamping diodes minimize switching transients and simplify system design.

These parallel-in or serial-in, serial-out shift registers have a complexity of 77 equivalent gates on a monolithic chip. They feature gated clock inputs and an overriding clear input. The parallel-in or serial-in modes are established by the shift/load input. When high, this input enables the serial data input and couples the eight flip-flops for serial shifting with each clock pulse. When low, the parallel (broadside) data inputs are enabled and synchronous loading occurs on the next clock pulse. During parallel loading, serial data flow is inhibited. Clocking is accomplished on the low-to-high-level edge of the clock pulse through a two-input positive NOR gate permitting one input to be used as a clock-enable or clock-inhibit function. Holding either of the clock inputs high inhibits clocking; holding either low enables the other clock input. This, of course, allows the system clock to be free-running and the register can be stopped on command with the other clock input. The clock inhibit input should be changed to the high level only while the clock input is high. A buffered, direct clear input overrides all other inputs, including the clock, and sets all flip-flops to zero.

SN54166, SN54LS166A . . . J OR W PACKAGE
SN74166 . . . N PACKAGE
SN74LS166A . . . D OR N PACKAGE
(TOP VIEW)

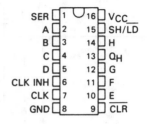

```
         ___
SER [ 1  U  16 ] Vcc
  A [ 2     15 ] SH/LD
  B [ 3     14 ] H
  C [ 4     13 ] Q_H
  D [ 5     12 ] G
CLK INH [ 6 11 ] F
CLK [ 7     10 ] E
GND [ 8      9 ] CLR
```

SN54LS166A . . . FK PACKAGE
(TOP VIEW)

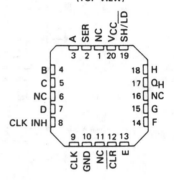

NC – No internal connection

logic symbol [†]

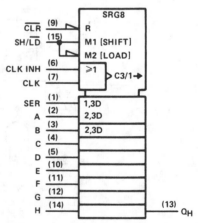

[†]This symbol is in accordance with ANSI/IEEE Std 91-1984 and IEC Publication 617-12.

Pin numbers shown are for D, J, N, and W packages.

FUNCTION TABLE

INPUTS						INTERNAL OUTPUTS		OUTPUT
CLEAR	SHIFT/LOAD	CLOCK INHIBIT	CLOCK	SERIAL	PARALLEL A . . . H	Q_A	Q_B	Q_H
L	X	X	X	X	X	L	L	L
H	X	L	L	X	X	Q_{A0}	Q_{B0}	Q_{H0}
H	L	L	↑	X	a . . . h	a	b	h
H	H	L	↑	H	X	H	Q_{An}	Q_{Gn}
H	H	L	↑	L	X	L	Q_{An}	Q_{Gn}
H	X	H	↑	X	X	Q_{A0}	Q_{B0}	Q_{H0}

TEXAS INSTRUMENTS
POST OFFICE BOX 655012 • DALLAS, TEXAS 75265

2-529

2

TTL Devices

SN54166, SN54LS166A, SN74166, SN74LS166A
PARALLEL-LOAD 8-BIT SHIFT REGISTERS

typical clear, shift, load, inhibit, and shift sequences

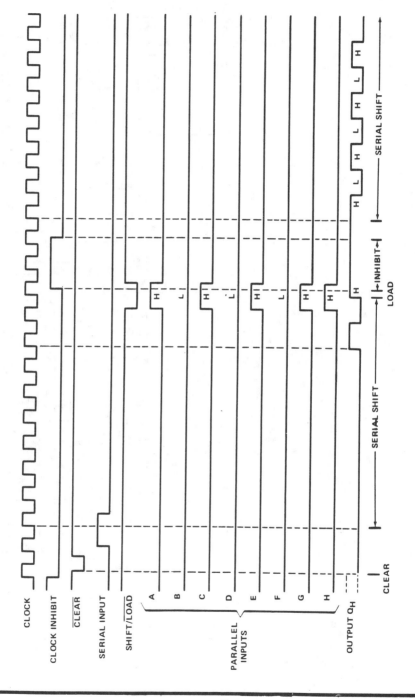

TEXAS
INSTRUMENTS
POST OFFICE BOX 655012 • DALLAS, TEXAS 75265

logic diagram (positive logic)

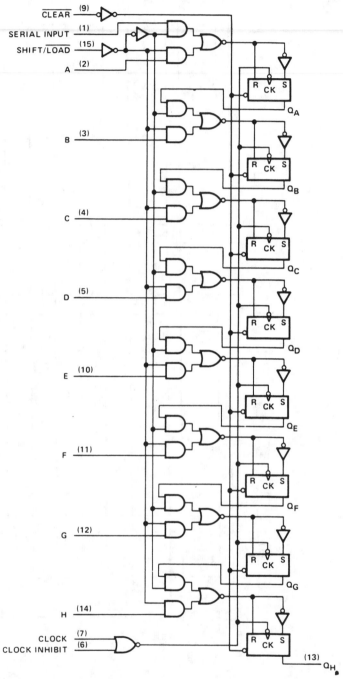

Pin numbers shown are for D, J, N, and W packages.

TEXAS
INSTRUMENTS
POST OFFICE BOX 655012 • DALLAS, TEXAS 75265

SN54184, SN74184 BCD-TO-BINARY CONVERTERS
SN54185A, SN74185A BINARY-TO-BCD CONVERTERS

SN54184, SN54185A . . . J OR W PACKAGE
SN74L184, SN74185A . . . N PACKAGE
(TOP VIEW)

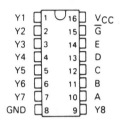

```
       ___
Y1  [ 1  U 16 ]  VCC
Y2  [ 2    15 ]  G
Y3  [ 3    14 ]  E
Y4  [ 4    13 ]  D
Y5  [ 5    12 ]  C
Y6  [ 6    11 ]  B
Y7  [ 7    10 ]  A
GND [ 8     9 ]  Y8
```

description

These monolithic converters are derived from the custom MSI 256-bit read-only memories SN5488 and SN7488. Emitter connections are made to provide direct read-out of converted codes at outputs Y8 through Y1 as shown in the function tables. These converters demonstrate the versatility of a read-only memory in that an unlimited number of reference tables or conversion tables may be built into a system using economical, customized read-only memories. Both of these converters comprehend that the least significant bits (LSB) of the binary and BCD codes are logically equal, and in each case the LSB bypasses the converter as illustrated in the typical applications. This means that a 6-bit converter is produced in each case. Both devices are cascadable to N bits.

TABLE I
SN54184, SN74184
PACKAGE COUNT AND DELAY TIMES
FOR BCD-TO-BINARY CONVERSION

INPUT (DECADES)	PACKAGES REQUIRED	TOTAL DELAY TIMES (ns)	
		TYP	MAX
2	2	56	80
3	6	140	200
4	11	196	280
5	19	280	400
6	28	364	520

An overriding enable input is provided on each converter which, when taken high, inhibits the function, causing all outputs to go high. For this reason, and to minimize power consumption, unused outputs Y7 and Y8 of the '185A and all "don't care" conditions of the '184 are programmed high. The outputs are of the open-collector type.

The SN54184 and SN54185A are characterized for operation over the full military temperature range of −55°C to 125°C; the SN74184 and SN74185A are characterized for operation from 0°C to 70°C.

SN54184 and SN74184 BCD-to-binary converters

The 6-bit BCD-to-binary function of the SN54184 and SN74184 is analogous to the algorithm:

a. Shift BCD number right one bit and examine each decade. Subtract three from each 4-bit decade containing a binary value greater than seven.

b. Shift right, examine, and correct after each shift until the least significant decade contains a number smaller than eight and all other converted decades contain zeros.

In addition to BCD-to-binary conversion, the SN54184 and SN74184 are programmed to generate BCD 9's complement or BCD 10's complement. Again, in each case, one bit of the complement code is logically equal to one of the BCD bits; therefore, these complements can be produced on three lines. As outputs Y6, Y7, and Y8 are not required in the BCD-to-binary conversion, they are utilized to provide these complement codes as specified in the function table (following page, right) when the devices are connected as shown above the function table.

TEXAS INSTRUMENTS

POST OFFICE BOX 655012 • DALLAS, TEXAS 75265

SN54184, SN74184
BCD-TO-BINARY AND BINARY-TO-BCD CONVERTERS

SN54184 and SN74184 BCD-to-binary converters (continued)

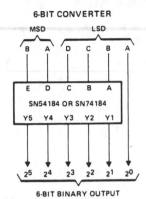

6-BIT CONVERTER

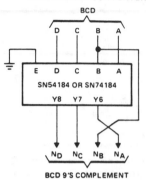

BCD 9'S COMPLEMENT CONVERTER

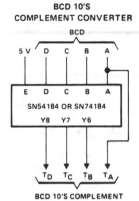

BCD 10'S COMPLEMENT CONVERTER

FUNCTION TABLE
BCD-TO-BINARY CONVERTER

BCD WORDS	INPUTS (See Note A)						OUTPUTS (See Note B)				
	E	D	C	B	A	$\overline{G}$	Y5	Y4	Y3	Y2	Y1
0-1	L	L	L	L	L	L	L	L	L	L	L
2-3	L	L	L	L	H	L	L	L	L	L	H
4-5	L	L	L	H	L	L	L	L	L	H	L
6-7	L	L	L	H	H	L	L	L	L	H	H
8-9	L	L	H	L	L	L	L	L	H	L	L
10-11	L	H	L	L	L	L	L	L	H	L	H
12-13	L	H	L	L	H	L	L	L	H	H	L
14-15	L	H	L	H	L	L	L	L	H	H	H
16-17	L	H	L	H	H	L	L	H	L	L	L
18-19	L	H	H	L	L	L	L	H	L	L	H
20-21	H	L	L	L	L	L	L	H	L	H	L
22-23	H	L	L	L	H	L	L	H	L	H	H
24-25	H	L	L	H	L	L	L	H	H	L	L
26-27	H	L	L	H	H	L	L	H	H	L	H
28-29	H	L	H	L	L	L	L	H	H	H	L
30-31	H	H	L	L	L	L	L	H	H	H	H
32-33	H	H	L	L	H	L	H	L	L	L	L
34-35	H	H	L	H	L	L	H	L	L	L	H
36-37	H	H	L	H	H	L	H	L	L	H	L
38-39	H	H	H	L	L	L	H	L	L	H	H
ANY	X	X	X	X	X	H	H	H	H	H	H

H = high level, L = low level, X = irrelevant

NOTES: A. Input conditions other than those shown produce highs at outputs Y1 through Y5.
  B. Outputs Y6, Y7, and Y8 are not used for BCD-to-binary conversion.

FUNCTION TABLE
BCD 9'S OR BCD 10'S COMPLEMENT CONVERTER

BCD WORD	INPUTS (See Note C)						OUTPUTS (See Note D)		
	E†	D	C	B	A	$\overline{G}$	Y8	Y7	Y6
0	L	L	L	L	L	L	H	L	H
1	L	L	L	L	H	L	H	L	L
2	L	L	L	H	L	L	L	H	H
3	L	L	L	H	H	L	L	H	L
4	L	L	H	L	L	L	L	H	H
5	L	L	H	L	H	L	L	H	L
6	L	L	H	H	L	L	L	L	H
7	L	L	H	H	H	L	L	L	L
8	L	H	L	L	L	L	L	L	L
9	L	H	L	L	H	L	L	L	L
0	H	L	L	L	L	L	L	L	L
1	H	L	L	L	H	L	H	L	L
2	H	L	L	H	L	L	H	L	L
3	H	L	L	H	H	L	L	H	H
4	H	L	H	L	L	L	L	H	H
5	H	L	H	L	H	L	L	L	H
6	H	L	H	H	L	L	L	H	L
7	H	L	H	H	H	L	L	L	L
8	H	H	L	L	L	L	L	L	H
9	H	H	L	L	H	L	L	L	L
ANY	X	X	X	X	X	H	H	H	H

H = high level, L = low level, X = irrelevant

NOTES: C. Input conditions other than those shown produce highs at outputs Y6, Y7, and Y8.
  D. Outputs Y1 through Y5 are not used for BCD 9's or BCD 10's complement conversion.

†When these devices are used as complement converters, input E is used as a mode control. With this input low, the BCD 9's complement is generated; when it is high, the BCD 10's complement is generated.

TEXAS INSTRUMENTS
POST OFFICE BOX 655012 • DALLAS, TEXAS 75265

SN54185A and SN74185A binary-to-BCD converters

The function performed by these 6-bit binary-to-BCD converters is analogous to the algorithm:

a. Examine the three most significant bits. If the sum is greater than four, add three and shift left one bit.

b. Examine each BCD decade. If the sum is greater than four, add three and shift left one bit.

c. Repeat step b until the least-significant binary bit is in the least-significant BCD location.

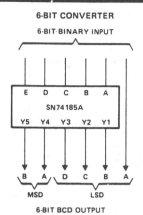

6-BIT CONVERTER

6-BIT BINARY INPUT

| E | D | C | B | A |
| SN74185A |
| Y5 | Y4 | Y3 | Y2 | Y1 |

B A | D C B A
MSD | LSD

6-BIT BCD OUTPUT

TABLE II

SN54185A, SN74185A

**PACKAGE COUNT AND DELAY TIMES
FOR BINARY-TO-BCD CONVERSION**

INPUT (BITS)	PACKAGES REQUIRED	TOTAL DELAY TIME (ns) TYP	TOTAL DELAY TIME (ns) MAX
4 to 6	1	25	40
7 or 8	3	50	80
9	4	75	120
10	6	100	160
11	7	125	200
12	8	125	200
13	10	150	240
14	12	175	280
15	14	175	280
16	16	200	320
17	19	225	360
18	21	225	360
19	24	250	400
20	27	275	440

FUNCTION TABLE

BINARY WORDS	INPUTS BINARY SELECT E	D	C	B	A	ENABLE G̅	OUTPUTS Y8	Y7	Y6	Y5	Y4	Y3	Y2	Y1
0 · 1	L	L	L	L	L	L	H	H	L	L	L	L	L	L
2 · 3	L	L	L	L	H	L	H	H	L	L	L	L	L	H
4 · 5	L	L	L	H	L	L	H	H	L	L	L	L	H	L
6 · 7	L	L	L	H	H	L	H	H	L	L	L	L	H	H
8 · 9	L	L	H	L	L	L	H	H	L	L	L	H	L	L
10 · 11	L	L	H	L	H	L	H	H	L	L	H	L	L	H
12 · 13	L	L	H	H	L	L	H	H	L	L	H	L	H	L
14 · 15	L	L	H	H	H	L	H	H	L	L	H	L	H	H
16 · 17	L	H	L	L	L	L	H	H	L	H	L	H	H	L
18 · 19	L	H	L	L	H	L	H	H	L	H	H	L	L	H
20 · 21	L	H	L	H	L	L	H	H	L	H	L	L	L	L
22 · 23	L	H	L	H	H	L	H	H	L	H	L	L	L	H
24 · 25	L	H	H	L	L	L	H	H	L	H	L	L	H	L
26 · 27	L	H	H	L	H	L	H	H	L	H	L	L	H	H
28 · 29	L	H	H	H	L	L	H	H	L	H	L	H	L	L
30 · 31	L	H	H	H	H	L	H	H	L	H	L	H	L	H
32 · 33	H	L	L	L	L	L	H	H	L	H	H	L	H	L
34 · 35	H	L	L	L	H	L	H	H	L	H	H	L	H	H
36 · 37	H	L	L	H	L	L	H	H	L	H	H	L	H	H
38 · 39	H	L	L	H	H	L	H	H	L	H	H	L	L	L
40 · 41	H	L	H	L	L	L	H	H	H	L	L	L	L	L
42 · 43	H	L	H	L	H	L	H	H	H	L	L	L	L	H
44 · 45	H	L	H	H	L	L	H	H	H	L	L	L	H	L
46 · 47	H	L	H	H	H	L	H	H	H	L	L	L	H	H
48 · 49	H	H	L	L	L	L	H	H	H	L	L	H	L	L
50 · 51	H	H	L	L	H	L	H	H	H	L	H	L	L	L
52 · 53	H	H	L	H	L	L	H	H	H	L	H	L	L	H
54 · 55	H	H	L	H	H	L	H	H	H	L	H	L	H	L
56 · 57	H	H	H	L	L	L	H	H	H	L	H	L	H	H
58 · 59	H	H	H	L	H	L	H	H	H	L	H	H	L	L
60 · 61	H	H	H	H	L	L	H	H	H	H	L	L	L	L
62 · 63	H	H	H	H	H	L	H	H	H	H	L	L	L	H
ALL	X	X	X	X	X	H	H	H	H	H	H	H	H	H

H = high level, L = low level, X = irrelevant

TEXAS INSTRUMENTS
POST OFFICE BOX 655012 • DALLAS, TEXAS 75265

SN54184, SN54185A, SN74184, SN74185A
BCD-TO-BINARY AND BINARY-TO-BCD CONVERTERS

logic symbols[†]

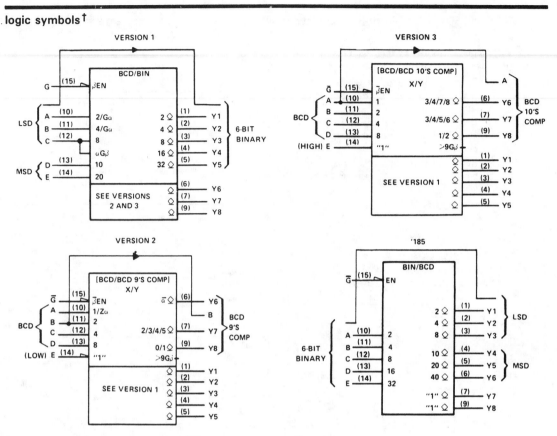

[†]These symbols are in accordance with ANSI/IEEE Std. 91-1984 and IEC Publication 617-12.

absolute maximum ratings over operating free-air temperature range (unless otherwise noted)

Supply voltage, V_{CC} (see Note 1) . . . . 7 V
Input voltage . . . . 5.5 V
Operating free-air temperature range: SN54184, SN54185A . . . . −55°C to 125°C
SN74184, SN74185A . . . . 0°C to 70°C
Storage temperature range . . . . −65°C to 150°C

NOTE 1: Voltage values are with respect to network ground terminal.

recommended operating conditions

	SN54184, SN54185A			SN74184, SN74185A			UNIT
	MIN	NOM	MAX	MIN	NOM	MAX	
Supply voltage, V_{CC}	4.5	5	5.5	4.75	5	5.25	V
Low-level output current, I_{OL}			12			12	mA
Operating free-air temperature, T_A	−55		125	0		70	C

Texas
INSTRUMENTS
POST OFFICE BOX 655012 • DALLAS, TEXAS 75265

SN54184, SN74184
BCD-TO-BINARY CONVERTERS

PARAMETER MEASUREMENT INFORMATION

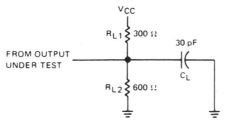

C_L includes probe and jig capacitance.

LOAD CIRCUIT
FIGURE 1

NOTE 2: Load circuits and voltage waveforms are shown in Section 1.

TYPICAL APPLICATION DATA
SN54184, SN74184

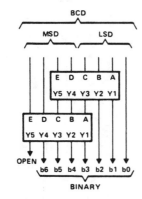

**FIGURE 2—BCD-TO-BINARY CONVERTER
FOR TWO BCD DECADES**

MSD—most significant decade
LSD—least significant decade
Each rectangle represents an SN54184 or SN74184

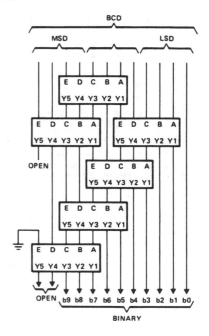

**FIGURE 3—BCD-TO-BINARY CONVERTER
FOR THREE BCD DECADES**

TEXAS
INSTRUMENTS
POST OFFICE BOX 655012 • DALLAS. TEXAS 75265

SN54185, SN74185A
BCD-TO-BINARY CONVERTERS

TYPICAL APPLICATION DATA
SN54185A, SN74185A

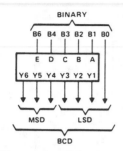

FIGURE 5—6-BIT BINARY-TO-BCD CONVERTER

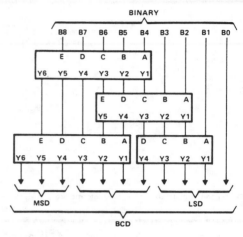

FIGURE 7—9-BIT BINARY-TO-BCD CONVERTER

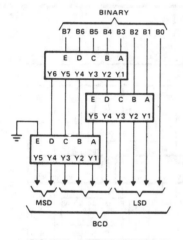

FIGURE 6—8-BIT BINARY-TO-BCD CONVERTER

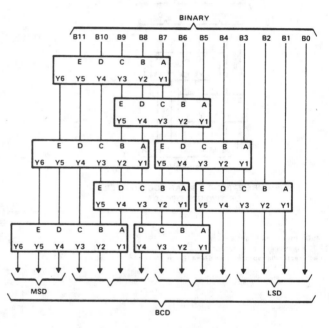

FIGURE 8—12-BIT BINARY-TO-BCD CONVERTER (SEE NOTE B)

MSD—Most significant decade
LSD—Least significant decade
NOTES: A. Each rectangle represents an SN54185A or an SN74185A.
        B. All unused E inputs are grounded.

TEXAS INSTRUMENTS
POST OFFICE BOX 655012 • DALLAS, TEXAS 75265

- **Counts 8-4-2-1 BCD or Binary**
- **Single Down/Up Count Control Line**
- **Count Enable Control Input**
- **Ripple Clock Output for Cascading**
- **Asynchronously Presettable with Load Control**
- **Parallel Outputs**
- **Cascadable for n-Bit Applications**

TYPE	AVERAGE PROPAGATION DELAY	TYPICAL MAXIMUM CLOCK FREQUENCY	TYPICAL POWER DISSIPATION
'190,'191	20ns	25MHz	325mW
'LS190,'LS191	20ns	25MHz	100mW

SN54190, SN54191, SN54LS190,
SN54LS191 . . . J PACKAGE
SN74190, SN74191 . . . N PACKAGE
SN74LS190, SN74LS191 . . . D OR N PACKAGE
(TOP VIEW)

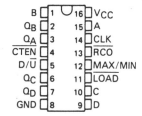

SN54LS190, SN54LS191 . . . FK PACKAGE
(TOP VIEW)

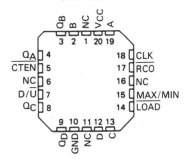

NC - No internal connection

description

The '190, 'LS190, '191, and 'LS191 are synchronous, reversible up/down counters having a complexity of 58 equivalent gates. The '191 and 'LS191 are 4-bit binary counters and the '190 and 'LS190 are BCD counters. Synchronous operation is provided by having all flip-flops clocked simultaneously so that the outputs change coincident with each other when so instructed by the steering logic. This mode of operation eliminates the output counting spikes normally associated with asynchronous (ripple clock) counters.

The outputs of the four master-slave flip-flops are triggered on a low-to-high transition of the clock input if the enable input is low. A high at the enable input inhibits counting. Level changes at the enable input should be made only when the clock input is high. The direction of the count is determined by the level of the down/up input. When low, the counter count up and when high, it counts down. A false clock may occur if the down/up input changes while the clock is low. A false ripple carry may occur if both the clock and enable are low and the down/up input is high during a load pulse.

These counters are fully programmable; that is, the outputs may be preset to either level by placing a low on the load input and entering the desired data at the data inputs. The output will change to agree with the data inputs independently of the level of the clock input. This feature allows the counters to be used as modulo-N dividers by simply modifying the count length with the preset inputs.

The clock, down/up, and load inputs are buffered to lower the drive requirement which significantly reduces the number of clock drivers, etc., required for long parallel words.

Two outputs have been made available to perform the cascading function: ripple clock and maximum/minimum count. The latter output produces a high-level output pulse with a duration approximately equal to one complete cycle of the clock when the counter overflows or underflows. The ripple clock output produces a low-level output pulse equal in width to the low-level portion of the clock input when an overflow or underflow condition exists. The counters can be easily cascaded by feeding the ripple clock output to the enable input of the succeeding counter if parallel clocking is used, or to the clock input if parallel enabling is used. The maximum/minimum count output can be used to accomplish look-ahead for high-speed operation.

Series 54' and 54LS' are characterized for operation over the full military temperature range of −55°C to 125°C; Series 74' and 74LS' are characterized for operation from 0°C to 70°C.

TEXAS
INSTRUMENTS

POST OFFICE BOX 655012 • DALLAS. TEXAS 75265

TTL Devices

2

logic symbols[†]

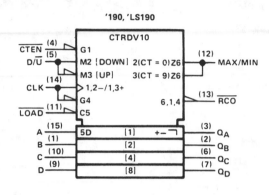

[†] These symbols are accordance with ANSI/IEEE Std 91-1984 and IEC Publication 617-12.
Pin numbers shown are for D, J, and N packages.

TEXAS
INSTRUMENTS

POST OFFICE BOX 655012 • DALLAS, TEXAS 75265

logic diagram (positive logic)

'190, 'LS190 DECADE COUNTERS

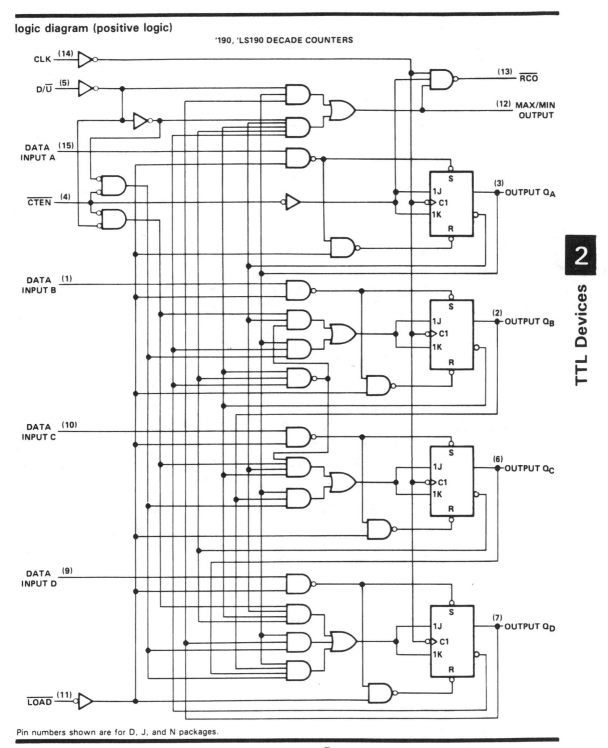

Pin numbers shown are for D, J, and N packages.

2

TTL Devices

TEXAS
INSTRUMENTS
POST OFFICE BOX 655012 • DALLAS, TEXAS 75265

2-621

logic diagram (positive logic)

'191, 'LS191 BINARY COUNTERS

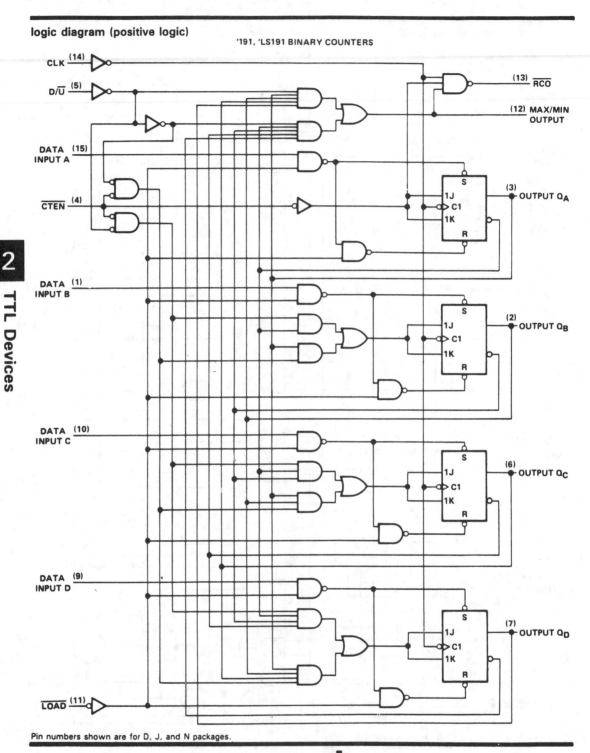

Pin numbers shown are for D, J, and N packages.

TEXAS INSTRUMENTS

POST OFFICE BOX 655012 • DALLAS. TEXAS 75265

'190, 'LS190 DECADE COUNTERS

typical load, count, and inhibit sequences

Illustrated below is the following sequence:

1. Load (preset) to BCD seven.
2. Count up to eight, nine (maximum), zero, one, and two.
3. Inhibit.
4. Count down to one, zero (minimum), nine, eight, and seven.

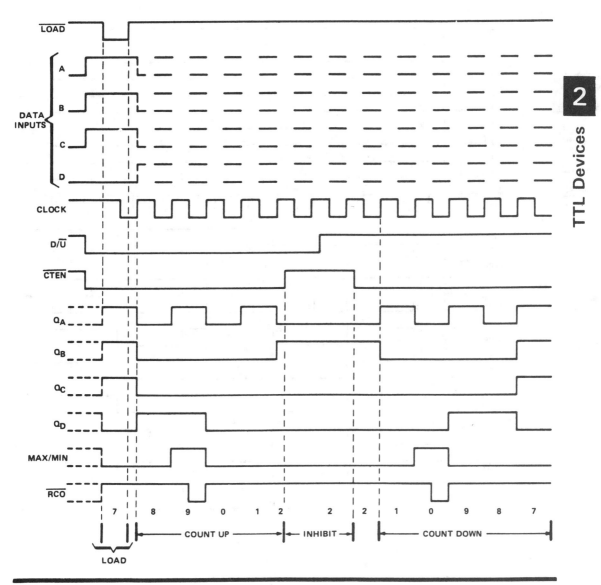

'191, 'LS191 BINARY COUNTERS

typical load, count, and inhibit sequences

Illustrated below is the following sequence:

1. Load (preset) to binary thirteen.
2. Count up to fourteen, fifteen (maximum), zero, one, and two.
3. Inhibit.
4. Count down to one, zero (minimum), fifteen, fourteen, and thirteen.

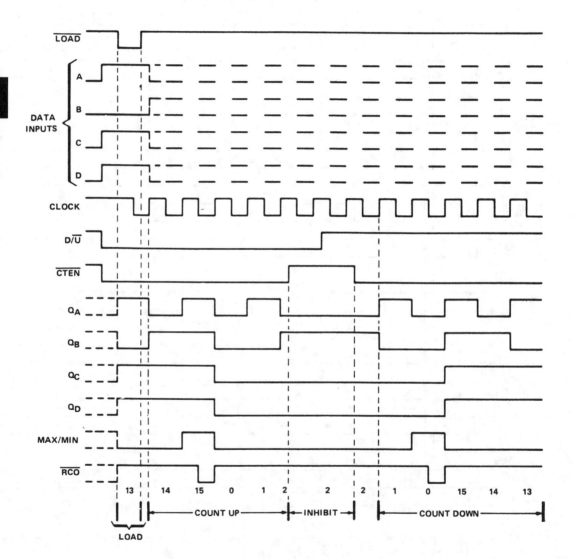

2

TTL Devices

TEXAS
INSTRUMENTS

POST OFFICE BOX 655012 • DALLAS, TEXAS 75265

SN54192, SN54193, SN54LS192 SN54LS193, SN74192, SN74193, SN74LS192, SN74LS193
SYNCHRONOUS 4-BIT UP/DOWN COUNTERS (DUAL CLOCK WITH CLEAR)

DECEMBER 1972 – REVISED MARCH 1988

- Cascading Circuitry Provided Internally
- Synchronous Operation
- Individual Preset to Each Flip-Flop
- Fully Independent Clear Input

TYPES	TYPICAL MAXIMUM COUNT FREQUENCY	TYPICAL POWER DISSIPATION
'192, '193	32 MHz	325 mW
'LS192, 'LS193	32 MHz	95 mW

SN54192, SN54193, SN54LS192,
SN54LS193 . . . J OR W PACKAGE
SN74192, SN74193 . . . N PACKAGE
SN74LS192, SN74LS193 . . . D OR N PACKAGE
(TOP VIEW)

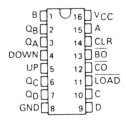

SN54LS192, SN54LS193 . . . FK PACKAGE
(TOP VIEW)

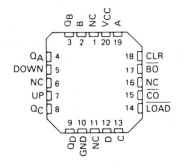

NC - No internal connection

description

These monolithic circuits are synchronous reversible (up/down) counters having a complexity of 55 equivalent gates. The '192 and 'LS192 circuits are BCD counters and the '193 and 'LS193 are 4-bit binary counters. Synchronous operation is provided by having all flip-flops clocked simultaneously so that the outputs change coincidently with each other when so instructed by the steering logic. This mode of operation eliminates the output counting spikes which are normally associated with asynchronous (ripple-clock) counters.

The outputs of the four master-slave flip-flops are triggered by a low-to-high-level transition of either count (clock) input. The direction of counting is determined by which count input is pulsed while the other count input is high.

All four counters are fully programmable; that is, each output may be preset to either level by entering the desired data at the data inputs while the load input is low. The output will change to agree with the data inputs independently of the count pulses. This feature allows the counters to be used as modulo-N dividers by simply modifying the count length with the preset inputs.

A clear input has been provided which forces all outputs to the low level when a high level is applied. The clear function is independent of the count and load inputs. The clear, count, and load inputs are buffered to lower the drive requirements. This reduces the number of clock drivers, etc., required for long words.

These counters were designed to be cascaded without the need for external circuitry. Both borrow and carry outputs are available to cascade both the up- and down-counting functions. The borrow output produces a pulse equal in width to the count-down input when the counter underflows. Similarly, the carry output produces a pulse equal in width to the count-up input when an overflow condition exists. The counters can then be easily cascaded by feeding the borrow and carry outputs to the count-down and count-up inputs respectively of the succeeding counter.

absolute maximum ratings over operating free-air temperature range (unless otherwise noted)

	SN54'	SN54LS'	SN74'	SN74LS'	UNIT
Supply voltage, V_{CC} (see Note 1)	7	7	7	7	V
Input voltage	5.5	7	5.5	7	V
Operating free-air temperature range	− 55 to 125		0 to 70		°C
Storage temperature range	− 65 to 150		− 65 to 150		°C

NOTE 1: Voltage values are with respect to network ground terminal.

TEXAS INSTRUMENTS
POST OFFICE BOX 655012 • DALLAS, TEXAS 75265

2

TTL Devices

'193, 'LS193 BINARY COUNTERS

typical clear, load, and count sequences

Illustrated below is the following sequence:

1. Clear outputs to zero.
2. Load (preset) to binary thirteen.
3. Count up to fourteen, fifteen, carry, zero, one, and two.
4. Count down to one, zero, borrow, fifteen, fourteen, and thirteen.

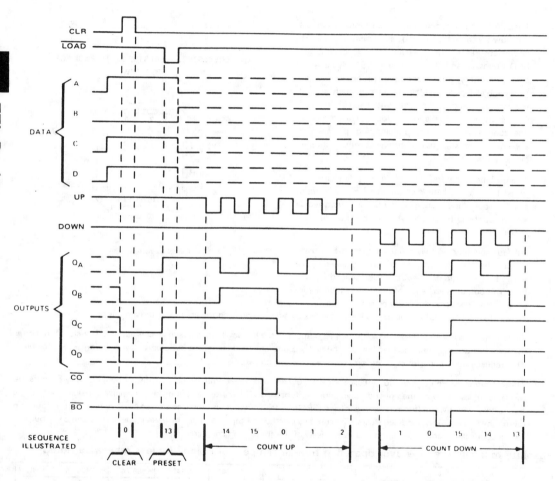

NOTES: A. Clear overrides load, data, and count inputs.

B. When counting up, count-down input must be high; when counting down, count-up input must be high.

TEXAS
INSTRUMENTS
POST OFFICE BOX 655012 • DALLAS, TEXAS 75265

- **SN54221, SN54LS221, SN74221 and SN74LS221 Are Dual Versions of Highly Stable SN54121, SN74121 One-Shots on a Monolithic Chip**

- **SN54221 and SN74221 Demonstrate Electrical and Switching Characteristics That Are Virtually Identical to the SN54121, SN74121 One-Shots**

- **Pin-Out Is Identical to the SN54123, SN74123, SN54LS123, SN74LS123**

- **Overriding Clear Terminates Output Pulse**

TYPE	TYPICAL POWER DISSIPATION,	MAXIMUM OUTPUT PULSE LENGTH
SN54221	130 mW	21 s
SN74221	130 mW	28 s
SN54LS221	23 mW	49 s
SN74LS221	23 mW	70 s

description

The '221 and 'LS221 are monolithic dual multi-vibrators with performance characteristics virtually identical to those of the '121. Each multivibrator features a negative-transition-triggered input and a positive-transition-triggered input either of which can be used as an inhibit input.

Pulse triggering occurs at a particular voltage level and is not directly related to the transition time of the input pulse. Schmitt-trigger input circuitry (TTL hysteresis) for B input allows jitter-free triggering from inputs with transition rates as slow as 1 volt/second, providing the circuit with excellent noise immunity of typically 1.2 volts. A high immunity to V_{CC} noise of typically 1.5 volts is also provided by internal latching circuitry.

Once fired, the outputs are independent of further transitions of the A and B inputs and are a function of the timing components, or the output pulses can be terminated by the overriding clear. Input pulses may be of any duration relative to the output pulse. Output pulse length may be varied from 35 nanoseconds to the maximums shown in the above table by choosing appropriate timing components. With R_{ext} = 2 kΩ and C_{ext} = 0, an output pulse of typically 30 nanoseconds is achieved which may be used as a d-c-triggered reset signal. Output rise and fall times are TTL compatible and independent of pulse length. Typical triggering and clearing sequences are illustrated as a part of the switching characteristics waveforms.

SN54221, SN54LS221 . . . J OR W PACKAGE
SN74221 . . . N PACKAGE
SN74LS221 . . . D OR N PACKAGE
(TOP VIEW)

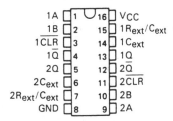

1A	1	16 V_{CC}
1B	2	15 $1R_{ext}/C_{ext}$
1$\overline{CLR}$	3	14 $1C_{ext}$
1$\overline{Q}$	4	13 1Q
2Q	5	12 2$\overline{Q}$
2C_{ext}	6	11 2$\overline{CLR}$
2R_{ext}/C_{ext}	7	10 2B
GND	8	9 2A

SN54LS221 . . . FK PACKAGE
(TOP VIEW)

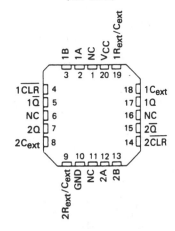

FUNCTION TABLE
(EACH MONOSTABLE)

INPUTS			OUTPUTS	
CLEAR	A	B	Q	$\overline{Q}$
L	X	X	L	H
X	H	X	L	H
X	X	L	L	H
H	L	↑	⊓ ‡	⊔ ‡
H	↓	H	⊓ ‡	⊔ ‡
↑	L	H	⊓ ‡	⊔ ‡
Also see description and switching characteristics				

†This condition is true only if the output of the latch formed by the two NAND gates has been conditioned to the logic 1 state prior to CLR going high. This latch is conditioned by taking either A high or B low while $\overline{CLR}$ is inactive (high).

‡Pulsed output patterns are tested during AC switching at 25 °C, with R_{ext} = 2 kΩ, C_{ext} = 80 pF.

TEXAS INSTRUMENTS

POST OFFICE BOX 655012 • DALLAS, TEXAS 75265

2

TTL Devices

description (continued)

Pulse width stability is achieved through internal compensation and is virtually independent of V_{CC} and temperature. In most applications, pulse stability will only be limited by the accuracy of external timing components.

Jitter-free operation is maintained over the full temperature and V_{CC} ranges for more than six decades of timing capacitance (10 pF to 10 μF) and more than one decade of timing resistance (2 kΩ to 30 kΩ for the SN54221, 2 kΩ to 40 kΩ for the SN74221, 2 kΩ to 70 kΩ for the SN54LS221, and 2 kΩ to 100 kΩ for the SN74LS221). Throughout these ranges, pulse width is defined by the relationship: $t_{w(out)} = C_{ext}R_{ext} \, ln2 \approx 0.7 \, C_{ext}R_{ext}$. In circuits where pulse cutoff is not critical, timing capacitance up to 1000 μF and timing resistance as low as 1.4 kΩ may be used. Also, the range of jitter-free output pulse widths is extended if V_{CC} is held to 5 volts and free-air temperature is 25°C. Duty cycles as high as 90% are achieved when using maximum recommended R_T. Higher duty cycles are available if a certain amount of pulse-width jitter is allowed.

The variance in output pulse width from device to device is typically less than ± 0.5% for given external timing components. An example of this distribution for the '221 is shown in Figure 2. Variations in output pulse width versus supply voltage and temperature for the '221 are shown in Figure 3 and 4, respectively.

Pin assignments for these devices are identical to those of the SN54123/SN74123 or SN54LS123/SN74LS123 so that the '221 or 'LS221 can be substituted for those products in systems not using the retrigger by merely changing the value of R_{ext} and/or C_{ext}, however the polarity of the capacitor will have to be changed.

TIMING COMPONENT CONNECTIONS

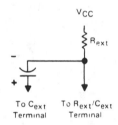

NOTE: Due to the internal circuit, the R_{ext}/C_{ext} pin will never be more positive than the C_{ext} pin.

Pin numbers shown are for D, J, N, and W packages.

logic symbol[†]

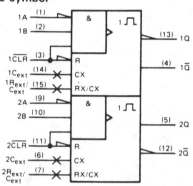

[†]This symbol is in accordance with ANSI/IEEE Std. 91-1984 and IEC Publication 617-12.

schematics of inputs and outputs

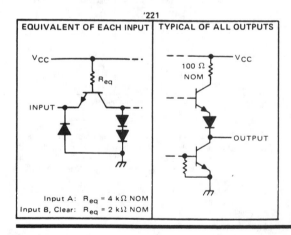

'221

EQUIVALENT OF EACH INPUT	TYPICAL OF ALL OUTPUTS

Input A: R_{eq} = 4 kΩ NOM
Input B, Clear: R_{eq} = 2 kΩ NOM

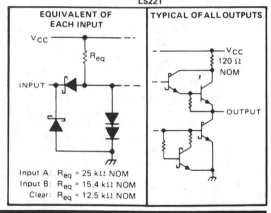

'LS221

EQUIVALENT OF EACH INPUT	TYPICAL OF ALL OUTPUTS

Input A: R_{eq} = 25 kΩ NOM
Input B: R_{eq} = 15.4 kΩ NOM
Clear: R_{eq} = 12.5 kΩ NOM

TEXAS
INSTRUMENTS
POST OFFICE BOX 655012 • DALLAS, TEXAS 75265

recommended operating conditions

		SN54LS221 MIN	NOM	MAX	SN74LS221 MIN	NOM	MAX	UNIT
Supply voltage, V_{CC}		4.5	5	5.5	4.75	5	5.25	V
High-level input voltage at A input, V_{IH}		2			2			V
Low-level input voltage at B input, V_{IL}				0.7			0.8	V
High-level output current, I_{OH}				-400			-400	μA
Low-level output current, I_{OL}				4			8	mA
Rate of rise or fall of input pulse, dv/dt	Schmitt, B	1			1			V/s
	Logic input, A	1			1			V/μs
Input pulse width	A or B, $t_{w(in)}$	50			50			ns
	Clear, $t_{w(clear)}$	40			40			ns
Clear-inactive-state setup time, t_{su}		15			15			ns
External timing resistance, R_{ext}		1.4		70	1.4		100	kΩ
External timing capacitance, C_{ext}		0		1000	0		1000	μF
Output duty cycle	R_T = 2 kΩ			50			50	%
	R_T = MAX R_{ext}			90			90	
Operating free-air temperature, T_A		-55		125	0		70	°C

recommended operating conditions

PARAMETER		TEST CONDITIONS[†]	SN54LS221 MIN	TYP[‡]	MAX	SN74LS221 MIN	TYP[‡]	MAX	UNIT
V_{T+}	Positive-going threshold voltage at B input	V_{CC} = MIN		1.0	2		1.0	2	V
V_{T-}	Negative-going threshold voltage at B input	V_{CC} = MIN	0.7	0.9		0.8	0.9		V
V_{IK}	Input clamp voltage	V_{CC} = MIN, I_I = -18 mA			-1.5			-1.5	V
V_{OH}	High-level output voltage	V_{CC} = MIN, I_{OH} = -400 μA	2.5	3.4		2.7	3.4		V
V_{OL}	Low-level output voltage	V_{CC} = MIN, I_{OL} = 4 mA		0.25	0.4		0.25	0.4	V
		I_{OL} = 8 mA					0.35	0.5	
I_I	Input current at maximum input voltage	V_{CC} = MAX, V_I = 7 V			0.1			0.1	mA
I_{IH}	High-level input current	V_{CC} = MAX, V_I = 2.7 V			20			20	μA
I_{IL}	Low-level input current	Input A	V_{CC} = MAX, V_I = 0.4 V			-0.4			-0.4	mA
		Input B			-0.8			-0.8	
		Clear			-0.8			-0.8	
I_{OS}	Short-circuit output current[§]	V_{CC} = MAX	-20		-100	-20		-100	mA
I_{CC}	Supply current	V_{CC} = MAX	Quiescent		4.7	11		4.7	11	mA
			Triggered		19	27		19	27	

[†]For conditions shown as MIN or MAX, use the appropriate value specified under recommended operating conditions.
[‡]All typical values are at V_{CC} = 5 V, T_A = 25°C.
[§]Not more than one output should be shorted at a time and duration of the short-circuit should not exceed one second.

2

TTL Devices

SN54LS221, SN74LS221
DUAL MONOSTABLE MULTIVIBRATORS
WITH SCHMITT-TRIGGER INPUTS

switching characteristics, $V_{CC} = 5$ V, $T_A = 25°C$

PARAMETER[†]	FROM (INPUT)	TO (OUTPUT)	TEST CONDITIONS		MIN	TYP	MAX	UNIT
t_{PLH}	A	Q	$C_L = 15$ pF, $R_L = 2$ kΩ, See Figure 1 and Note 3	$C_{ext} = 80$ pF, $R_{ext} = 2$ kΩ		45	70	ns
	B	Q				35	55	
t_{PHL}	A	$\overline{Q}$				50	80	ns
	B	$\overline{Q}$				40	65	
t_{PHL}	Clear	Q				35	55	ns
t_{PLH}	Clear	$\overline{Q}$				44	65	ns
$t_{w(out)}$	A or B	Q or $\overline{Q}$		$C_{ext} = 80$ pF, $R_{ext} = 2$ kΩ	70	120	150	ns
				$C_{ext} = 0$, $R_{ext} = 2$ kΩ	20	47	70	
				$C_{ext} = 100$ pF, $R_{ext} = 10$ kΩ	670	740	810	
				$C_{ext} = 1$ μF, $R_{ext} = 10$ kΩ	6	6.9	7.5	ms

[†]t_{PLH} = Propagation delay time, low-to-high-level output
t_{PHL} = Propagation delay time, high-to-low-level output
$t_{w(out)}$ = Output pulse width
NOTE 3: Load circuits and voltage waveforms are shown in Section 1.

TEXAS
INSTRUMENTS

POST OFFICE BOX 655012 • DALLAS, TEXAS 75265

TYPICAL CHARACTERISTICS ('221 ONLY)†

TTL Devices

2

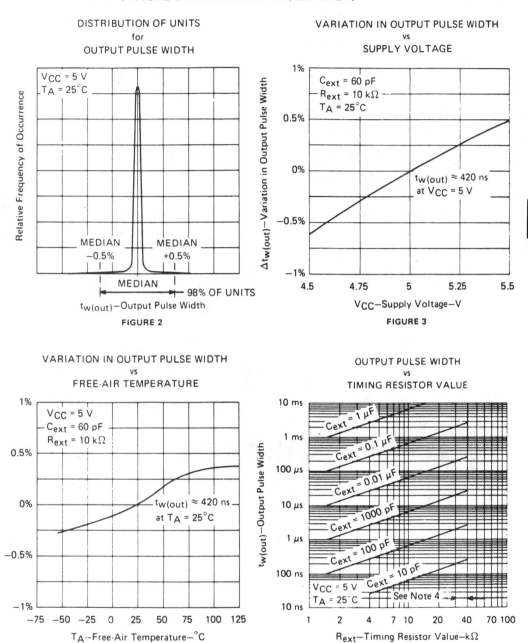

DISTRIBUTION OF UNITS
for
OUTPUT PULSE WIDTH

FIGURE 2

VARIATION IN OUTPUT PULSE WIDTH
vs
SUPPLY VOLTAGE

FIGURE 3

VARIATION IN OUTPUT PULSE WIDTH
vs
FREE-AIR TEMPERATURE

FIGURE 4

OUTPUT PULSE WIDTH
vs
TIMING RESISTOR VALUE

FIGURE 5

NOTE 4: These values of resistance exceed the maximum recommended for use over the full temperature range of the SN54221.

†Data for temperatures below 0 C and above 70°C, and for supply voltages below 4.75 V and above 5.25 V are applicable for the SN54221 only.

TEXAS INSTRUMENTS

POST OFFICE BOX 655012 • DALLAS. TEXAS 75265

SN54LS240, SN54LS241, SN54LS244, SN54S240, SN54S241, SN54S244, SN74LS240, SN74LS241, SN74LS244, SN74S240, SN74S241, SN74S244
OCTAL BUFFERS AND LINE DRIVERS WITH 3-STATE OUTPUTS

APRIL 1985 – REVISED MARCH 1988

- **3-State Outputs Drive Bus Lines or Buffer Memory Address Registers**
- **PNP Inputs Reduce D-C Loading**
- **Hysteresis at Inputs Improves Noise Margins**

description

These octal buffers and line drivers are designed specifically to improve both the performance and density of three-state memory address drivers, clock drivers, and bus-oriented receivers and transmitters. The designer has a choice of selected combinations of inverting and noninverting outputs, symmetrical $\overline{G}$ (active-low output control) inputs, and complementary G and $\overline{G}$ inputs. These devices feature high fan-out, improved fan-in, and 400-mV noise-margin. The SN74LS' and SN74S' can be used to drive terminated lines down to 133 ohms.

The SN54' family is characterized for operation over the full military temperature range of −55°C to·125°C. The SN74' family is characterized for operation from 0°C to 70°C.

SN54LS', SN54S' . . . J OR W PACKAGE
SN74LS', SN74S' . . . DW OR N PACKAGE
(TOP VIEW)

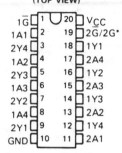

1$\overline{G}$	1	20	V$_{CC}$
1A1	2	19	2G/2$\overline{G}$*
2Y4	3	18	1Y1
1A2	4	17	2A4
2Y3	5	16	1Y2
1A3	6	15	2A3
2Y2	7	14	1Y3
1A4	8	13	2A2
2Y1	9	12	1Y4
GND	10	11	2A1

SN54LS', SN54S' . . . FK PACKAGE
(TOP VIEW)

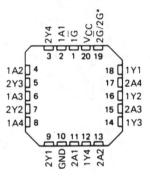

*2G for 'LS241 and 'S241 or 2$\overline{G}$ for all other drivers.

schematics of inputs and outputs

'LS240, 'LS241, 'LS244	'S240, 'S241, 'S244	
EQUIVALENT OF EACH INPUT	EQUIVALENT OF EACH INPUT	TYPICAL OF ALL OUTPUTS

G and $\overline{G}$ inputs: R$_{eq}$ = 2 kΩ NOM
A inputs: R$_{eq}$ = 2.8 kΩ NOM

'LS240, 'LS241, 'LS244;
R = 50 Ω NOM
'S240, 'S241, 'S244
R = 25 Ω NOM

TEXAS INSTRUMENTS
POST OFFICE BOX 655012 • DALLAS, TEXAS 75265

2

TTL Devices

SN54LS240, SN54LS241, SN54LS244, SN54S240, SN54S241, SN54S244, SN74SL240, SN74LS241, SN74LS244, SN74S240, SN74S241, SN74S244
OCTAL BUFFERS AND LINE DRIVERS WITH 3-STATE OUTPUTS

logic symbols†

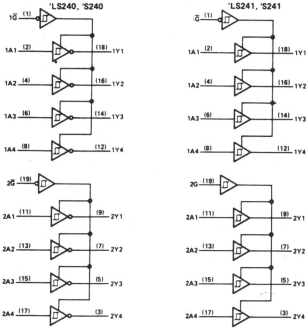
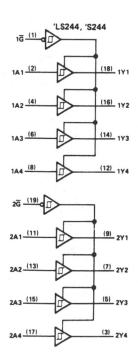

†These symbols are in accordance with ANSI/IEEE Std. 91-1984 and IEC Publication 617-12.

logic diagrams (positive logic)

Pin numbers shown are for DW, J, N, and W packages.

absolute maximum ratings over operating free-air temperature range (unless otherwise noted)

Supply voltage, V_{CC} (see Note 1). . . . . 7 V
Input voltage: 'LS Circuits. . . . . 7 V
'S Circuits. . . . . 5.5 V
Off-state output voltage . . . . 5.5 V
Operating free-air temperature range: SN54LS', SN54S' Circuits . . . . −55°C to 125°C
SN74LS', SN74S' Circuits . . . . 0°C to 70°C
Storage temperature range . . . . − 65°C to 150°C

NOTE 1: Voltage values are with respect to network ground terminal.

TEXAS INSTRUMENTS

POST OFFICE BOX 655012 • DALLAS. TEXAS 75265

2

TTL Devices

- **Package Options Include Ceramic Chip Carriers and Flat Packages in Addition to Plastic and Ceramic DIPs**

- **Dependable Texas Instruments Quality and Reliability**

SN54S260 . . . J OR W PACKAGE
SN74S260 . . . D OR N PACKAGE
(TOP VIEW)

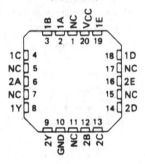

1A	1	14	V$_{CC}$
1B	2	13	1E
1C	3	12	1D
2A	4	11	2E
1Y	5	10	2D
2Y	6	9	2C
GND	7	8	2B

description

These devices contain two independent 5-input positive -NOR gates. They perform the Boolean function $Y = \overline{A + B + C + D + E}$ in positive logic.

The SN54S260 is characterized for operation over the full military temperature range of $-55°C$ to $125°C$. The SN74S260 is characterized for operation from $0°C$ to $70°C$.

SN54S260 . . . FK PACKAGE
(TOP VIEW)

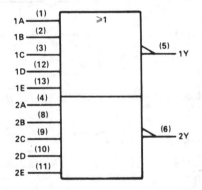

NC - No internal connection

logic diagram (each gate)

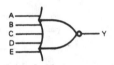

logic symbol†

```
        (1)   ┌────────────┐
1A ─────────  │    ≥1      │
        (2)   │            │
1B ─────────  │            │
        (3)   │            │      (5)
1C ─────────  │            │──────╲──── 1Y
       (12)   │            │
1D ─────────  │            │
       (13)   │            │
1E ─────────  │            │
              ├────────────┤
        (4)   │            │
2A ─────────  │            │
        (8)   │            │
2B ─────────  │            │
        (9)   │            │      (6)
2C ─────────  │            │──────╲──── 2Y
       (10)   │            │
2D ─────────  │            │
       (11)   │            │
2E ─────────  └────────────┘
```

†This symbol is in accordance with ANSI/IEEE Std. 91-1984 and IEC Publication 617-12.
Pin numbers shown are for D, J, N, and W packages.

TEXAS INSTRUMENTS
POST OFFICE BOX 655012 • DALLAS, TEXAS 75265

TTL Devices

2

SN54LS373, SN54LS374, SN54S373, SN54S374, SN74LS373, SN74LS374, SN74S373, SN74S374
OCTAL D-TYPE TRANSPARENT LATCHES AND EDGE-TRIGGERED FLIP-FLOPS

OCTOBER 1975 — REVISED MARCH 1988

- Choice of 8 Latches or 8 D-Type Flip-Flops In a Single Package

- 3-State Bus-Driving Outputs

- Full Parallel-Access for Loading

- Buffered Control Inputs

- Clock/Enable Input Has Hysteresis to Improve Noise Rejection ('S373 and 'S374)

- P-N-P Inputs Reduce D-C Loading on Data Lines ('S373 and 'S374)

SN54LS373, SN54LS374, SN54S373,
SN54S374 . . . J OR W PACKAGE
SN74LS373, SN74LS374, SN74S373,
SN74S374 . . . DW OR N PACKAGE
(TOP VIEW)

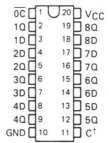

SN54LS373, SN54LS374, SN54S373,
SN54S374 . . . FK PACKAGE
(TOP VIEW)

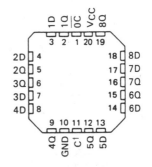

†C for 'LS373 and 'S373; CLK for 'LS374 and 'S374.

'LS373, 'S373 FUNCTION TABLE

OUTPUT ENABLE	ENABLE LATCH	D	OUTPUT
L	H	H	H
L	H	L	L
L	L	X	Q_0
H	X	X	Z

'LS374, 'S374 FUNCTION TABLE

OUTPUT ENABLE	CLOCK	D	OUTPUT
L	↑	H	H
L	↑	L	L
L	L	X	Q_0
H	X	X	Z

description

These 8-bit registers feature three-state outputs designed specifically for driving highly-capacitive or relatively low-impedance loads. The high-impedance third state and increased high-logic-level drive provide these registers with the capability of being connected directly to and driving the bus lines in a bus-organized system without need for interface or pull-up components. They are particularly attractive for implementing buffer registers, I/O ports, bidirectional bus drivers, and working registers.

The eight latches of the 'LS373 and 'S373 are transparent D-type latches meaning that while the enable (C) is high the Q outputs will follow the data (D) inputs. When the enable is taken low the output will be latched at the level of the data that was set up.

2 TTL Devices

TEXAS
INSTRUMENTS

POST OFFICE BOX 655012 • DALLAS, TEXAS 75265

SN54LS373, SN54LS374, SN54S373, SN54S374,
SN74LS373, SN74LS374, SN74S373, SN74S374
OCTAL D-TYPE TRANSPARENT LATCHES AND EDGE-TRIGGERED FLIP-FLOPS

description (continued)

The eight flip-flops of the 'LS374 and 'S374 are edge-triggered D-type flip-flops. On the positive transition of the clock, the Q outputs will be set to the logic states that were setup at the D inputs.

Schmitt-trigger buffered inputs at the enable/clock lines of the 'S373 and 'S374 devices, simplify system design as ac and dc noise rejection is improved by typically 400 mV due to the input hysteresis. A buffered output control input can be used to place the eight outputs in either a normal logic state (high or low logic levels) or a high-impedance state. In the high-impedance state the outputs neither load nor drive the bus lines significantly.

The output control does not affect the internal operation of the latches or flip-flops. That is, the old data can be retained or new data can be entered even while the outputs are off.

logic diagrams (positive logic)

'LS373, 'S373
TRANSPARENT LATCHES

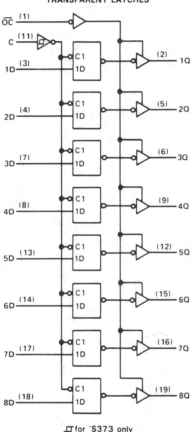

'LS374, 'S374
POSITIVE-EDGE-TRIGGERED FLIP-FLOPS

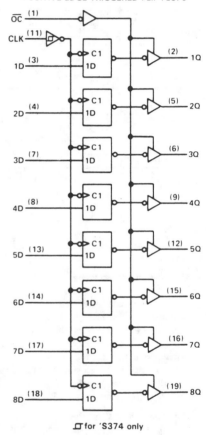

⊓ for 'S373 only   ⊓ for 'S374 only

Pin numbers shown are for DW, J, N, and W packages.

TEXAS
INSTRUMENTS
POST OFFICE BOX 655012 • DALLAS, TEXAS 75265

SN54390, SN54LS390, SN54393, SN54LS393, SN74390, SN74LS390, SN74393, SN74LS393
DUAL 4-BIT DECADE AND BINARY COUNTERS

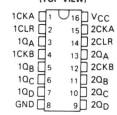

OCTOBER 1976 — REVISED MARCH 1988

- Dual Versions of the Popular '90A, 'LS90 and '93A, 'LS93

- '390, 'LS390 . . . Individual Clocks for A and B Flip-Flops Provide Dual ÷ 2 and ÷ 5 Counters

- '393, 'LS393 . . . Dual 4-Bit Binary Counter with Individual Clocks

- All Have Direct Clear for Each 4-Bit Counter

- Dual 4-Bit Versions Can Significantly Improve System Densities by Reducing Counter Package Count by 50%

- Typical Maximum Count Frequency . . . 35 MHz

- Buffered Outputs Reduce Possibility of Collector Commutation

description

Each of these monolithic circuits contains eight master-slave flip-flops and additional gating to implement two individual four-bit counters in a single package. The '390 and 'LS390 incorporate dual divide-by-two and divide-by-five counters, which can be used to implement cycle lengths equal to any whole and/or cumulative multiples of 2 and/or 5 up to divide-by-100. When connected as a bi-quinary counter, the separate divide-by-two circuit can be used to provide symmetry (a square wave) at the final output stage. The '393 and 'LS393 each comprise two independent four-bit binary counters each having a clear and a clock input. N-bit binary counters can be implemented with each package providing the capability of divide-by-256. The '390, 'LS390, '393, and 'LS393 have parallel outputs from each counter stage so that any submultiple of the input count frequency is available for system-timing signals.

Series 54 and Series 54LS circuits are characterized for operation over the full military temperature range of −55°C to 125°C; Series 74 and Series 74LS circuits are characterized for operation from 0°C to 70°C.

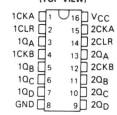

SN54390, SN54LS390 . . . J OR W PACKAGE
SN74390 . . . N PACKAGE
SN74LS390 . . . D OR N PACKAGE
(TOP VIEW)

1CKA	1	16	VCC
1CLR	2	15	2CKA
1QA	3	14	2CLR
1CKB	4	13	2QA
1QB	5	12	2CKB
1QC	6	11	2QB
1QD	7	10	2QC
GND	8	9	2QD

SN54LS390 . . . FK PACKAGE
(TOP VIEW)

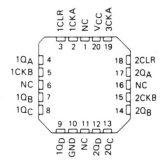

SN54393, SN54LS393 . . . J OR W PACKAGE
SN74393 . . . N PACKAGE
SN74LS393 . . . D OR N PACKAGE
(TOP VIEW)

1A	1	14	VCC
1CLR	2	13	2A
1QA	3	12	2CLR
1QB	4	11	2QA
1QC	5	10	2QB
1QD	6	9	2QC
GND	7	8	2QD

SN54LS393 . . . FK PACKAGE
(TOP VIEW)

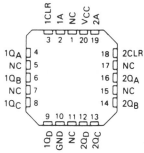

NC - No internal connection

2 TTL Devices

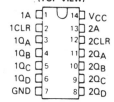

TEXAS INSTRUMENTS
POST OFFICE BOX 655012 • DALLAS, TEXAS 75265

SN54390, SN54LS390, SN54393, SN54LS393, SN74390, SN74LS390, SN74393, SN74LS393
DUAL 4-BIT DECADE AND BINARY COUNTERS

FUNCTION TABLES

'390, 'LS390
BCD COUNT SEQUENCE
(EACH COUNTER)
(See Note A)

COUNT	OUTPUT			
	Q_D	Q_C	Q_B	Q_A
0	L	L	L	L
1	L	L	L	H
2	L	L	H	L
3	L	L	H	H
4	L	H	L	L
5	L	H	L	H
6	L	H	H	L
7	L	H	H	H
8	H	L	L	L
9	H	L	L	H

'390, 'LS390
BI-QUINARY (5-2)
(EACH COUNTER)
(See Note B)

COUNT	OUTPUT			
	Q_A	Q_D	Q_C	Q_B
0	L	L	L	L
1	L	L	L	H
2	L	L	H	L
3	L	L	H	H
4	L	H	L	L
5	H	L	L	L
6	H	L	L	H
7	H	L	H	L
8	H	L	H	H
9	H	H	L	L

'393, 'LS393
COUNT SEQUENCE
(EACH COUNTER)

COUNT	OUTPUT			
	Q_D	Q_C	Q_B	Q_A
0	L	L	L	L
1	L	L	L	H
2	L	L	H	L
3	L	L	H	H
4	L	H	L	L
5	L	H	L	H
6	L	H	H	L
7	L	H	H	H
8	H	L	L	L
9	H	L	L	H
10	H	L	H	L
11	H	L	H	H
12	H	H	L	L
13	H	H	L	H
14	H	H	H	L
15	H	H	H	H

NOTES: A. Output Q_A is connected to input B for BCD count.
B. Output Q_D is connected to input A for bi-quinary count.
C. H = high level, L = low level.

logic diagrams (positive logic)

'390, 'LS390

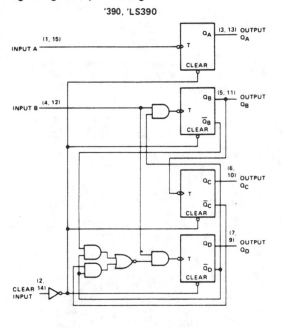

logic symbols†

'390, 'LS390

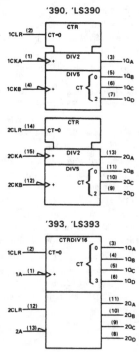

'393, 'LS393

†These symbols are in accordance with ANSI/IEEE Std. 91-1984 and IEC Publication 617-12.

Pin numbers shown are for D, J, N, and W packages.

TEXAS INSTRUMENTS
POST OFFICE BOX 655012 • DALLAS, TEXAS 75265

logic diagrams (continued)

'393, 'LS393

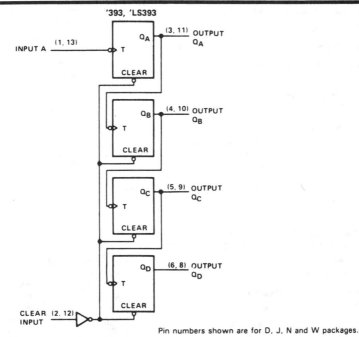

Pin numbers shown are for D, J, N and W packages.

schematics of inputs and outputs

'390, '393

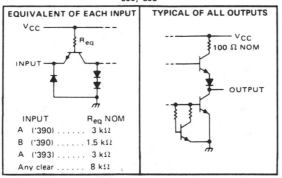

'LS390, 'LS393

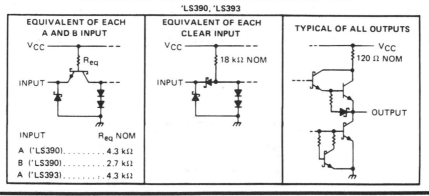

TEXAS
INSTRUMENTS
POST OFFICE BOX 655012 • DALLAS, TEXAS 75265

- Previously Called TMS4045/TMS40L45
- 1024 X 4 Organization
- Single +5-V Supply
- High Density 300-mil (7.62 mm) 18-Pin Package
- Fully Static Operation (No Clocks, No Refresh, No Timing Strobe)
- 4 Performance Ranges:

	ACCESS READ OR WRITE	
	TIME (MAX)	CYCLE (MIN)
TMS2114-15, TMS2114L-15	150 ns	150 ns
TMS2114-20, TMS2114L-20	200 ns	200 ns
TMS2114-25, TMS2114L-25	250 ns	250 ns
TMS2114-45, TMS2114L-45	450 ns	450 ns

- 400-mV Guaranteed DC Noise Immunity with Standard TTL Loads – No Pull-Up Resistors Required
- Common I/O Capability
- 3-State Outputs and Chip Select Control for OR-Tie Capability
- Fan-Out to 2 Series 74, 1 Series 74S, or 8 Series 74LS TTL Loads
- Low Power Dissipation

	MAX (OPERATING)
TMS2114	550 mW
TMS2114L	330 mW

TMS2114, TMS2114L . . . NL PACKAGE
(TOP VIEW)

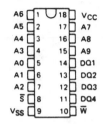

A6	1	18	VCC
A5	2	17	A7
A4	3	16	A8
A3	4	15	A9
A0	5	14	DQ1
A1	6	13	DQ2
A2	7	12	DQ3
$\overline{S}$	8	11	DQ4
VSS	9	10	$\overline{W}$

PIN NOMENCLATURE	
A0 – A9	Addresses
DQ1 – DQ4	Data In/Data Out
$\overline{S}$	Chip Select
VCC	+5-V Supply
VSS	Ground
$\overline{W}$	Write Enable

Static RAM and Memory Support Devices

8

description

This series of static random-access memories is organized as 1024 words of 4 bits each. Static design results in reducing overhead costs by elimination of refresh-clocking circuitry and by simplification of timing requirements. Because this series is fully static, chip select may be tied low to further simplify system timing. Output data is always available during a read cycle.

All inputs and outputs are fully compatible with Series 74, 74S or 74LS TTL. No pull-up resistors are required. This 4K Static RAM series is manufactured using TI's reliable N-channel silicon-gate technology to optimize the cost/performance relationship.

The TMS2114/2114L series is offered in the 18-pin dual-in-line plastic (NL suffix) package designed for insertion in mounting-hole rows on 300-mil (7.62 mm) centers. The series is guaranteed for operation from 0°C to 70°C.

TEXAS
INSTRUMENTS
POST OFFICE BOX 225012 • DALLAS, TEXAS 75265

operation

addresses (A0 – A9)

The ten address inputs select one of the 1024 4-bit words in the RAM. The address inputs must be stable for the duration of a write cycle. The address inputs can be driven directly from standard Series 54/74 TTL with no external pull-up resistors.

chip select ($\overline{S}$)

The chip-select terminal, which can be driven directly from standard TTL circuits, affects the data-in and data-out terminals. When chip select is at a logic low level, both terminals are enabled. When chip select is high, data-in is inhibited and data-out is in the floating or high-impedance state.

write enable ($\overline{W}$)

The read or write mode is selected through the write enable terminal. A logic high selects the read mode; a logic low selects the write mode. $\overline{W}$ or $\overline{S}$ must be high when changing addresses to prevent erroneously writing data into a memory location. The $\overline{W}$ input can be driven directly from standard TTL circuits.

data-in/data-out (DQ1 – DQ4)

Data can be written into a selected device when the write enable input is low. The DQ terminal can be driven directly from standard TTL circuits. The three-state output buffer provides direct TTL compatibility with a fan-out of two Series 74 TTL gates, one Series 74S TTL gate, or eight Series 74LS TTL gates. The DQ terminals are in the high-impedance state when chip select ($\overline{S}$) is high or whenever a write operation is being performed. Data-out is the same polarity as data-in.

Static RAM and Memory Support Devices

8

TEXAS INSTRUMENTS
POST OFFICE BOX 225012 ● DALLAS, TEXAS 75265

logic symbol†

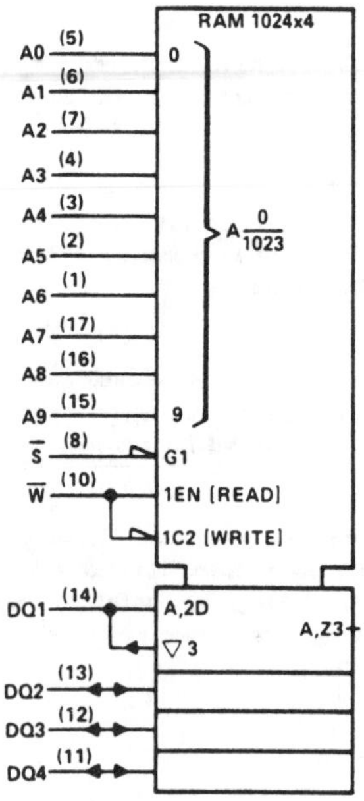

FUNCTION TABLE

$\overline{W}$	$\overline{S}$	DQ1 – DQ4	MODE
L	L	VALID DATA	WRITE
H	L	DATA OUTPUT	READ
X	H	HI-Z	DEVICE DISABLED

†This symbol is in accordance with IEEE Std 91/ANSI Y32.14 and recent decisions by IEEE and IEC. See explanation on page 10-1.

absolute maximum ratings over operating free-air temperature (unless otherwise noted)†

Supply voltage, V_{CC} (see Note 1) . . . . −0.5 V to 7 V
Input voltage (any input) (see Note 1) . . . . −1 V to 7 V
Continuous power dissipation . . . . 1 W
Operating free-air temperature range . . . . 0°C to 70°C
Storage temperature range . . . . −55°C to 150°C

Stresses beyond those listed under "Absolute Maximum Rating" may cause permanent damage to the device. This is a stress rating only and functional operation of the device at these or any other conditions beyond those indicated in the "Recommended Operating Conditions" section of this specification is not implied. Exposure to absolute-maximum-rated conditions for extended periods may affect device reliability.

NOTE 1: Voltage values are with respect to the ground material.

recommended operating conditions

PARAMETER	TMS2114 TMS2114L MIN	NOM	MAX	UNIT
Supply voltage, V_{CC}	4.5	5	5.5	V
Supply voltage, V_{SS}		0		V
High-level input voltage, V_{IH}	2		5.5	V
Low-level input voltage, V_{IL} (see Note 2)	−1		0.8	V
Operating free-air temperature, T_A	0		70	°C

NOTE 2: The algebraic convention, where the more negative (less positive) limit is designated as minium, is used in this data sheet for logic voltage levels only.

electrical characteristics over recommended operating free-air temperature range (unless otherwise noted)

	PARAMETER	TEST CONDITIONS[†]			MIN	TYP[‡]	MAX	UNIT
V_{OH}	High-level voltage	$I_{OH} = -1$ mA	V_{CC} = MIN (operating)		2.4			V
V_{OL}	Low-level voltage	$I_{OL} = 3.2$ mA	V_{CC} = MIN (operating)				0.4	V
I_I	Input current	$V_I = 0$ V to MAX					10	μA
I_{OZ}	Off-state output current	$\overline{S}$ at 2 V or $\overline{W}$ at 0.8 V	$V_O = 0$ V to MAX				±10	μA
I_{CC}	Supply current from V_{CC}	$I_O = 0$ mA, $T_A = 0°$C (worst case)	TMS 2114	V_{CC} = MAX		90	100	mA
			TMS 2114L	V_{CC} = MAX		50	60	
C_i	Input capacitance	$V_I = 0$ V, f = 1 MHz					8	pF
C_o	Output capacitance	$V_O = 0$ V, f = 1 MHz					8	pF

[†] For conditions shown as MIN or MAX, use the appropriate value specified under recommended operating conditions.
[‡] All typical values are at V_{CC} = 5 V, $T_A = 25°$C.

timing requirements over recommended supply voltage range, $T_A = 0°$C to 70°C, 1 Series 74 TTL load, $C_L = 100$ pF

	PARAMETER	TMS2114-15 TMS2114L-15		TMS2114-20 TMS2114L-20		TMS2114-25 TMS2114L-25		TMS2114-45 TMS2114L-45		UNIT
		MIN	MAX	MIN	MAX	MIN	MAX	MIN	MAX	
$t_{c(rd)}$	Read cycle time	150		200		250		450		ns
$t_{c(wr)}$	Write cycle time	150		200		250		450		ns
$t_{w(W)}$	Write pulse width	80		100		100		200		ns
$t_{su(A)}$	Address set up time	0		0		0		0		ns
$t_{su(S)}$	Chip select set up time	80		100		100		200		ns
$t_{su(D)}$	Data set up time	80		100		100		200		ns
$t_{h(D)}$	Data hold time	0		0		0		0		ns
$t_{h(A)}$	Address hold time	0		0		0		20		ns

switching characteristics over recommended voltage range, T_A = 0°C to 70°C, 1 Series 74 TTL load, C_L = 100 pF

	PARAMETER	TMS2114-15 TMS2114L-15		TMS2114-20 TMS2114L-20		TMS2114-25 TMS2114L-25		TMS2114-45 TMS2114L-45		UNIT
		MIN	MAX	MIN	MAX	MIN	MAX	MIN	MAX	
$t_{a(A)}$	Access time from address		150		200		250		450	ns
$t_{a(S)}$	Access time from chip select (or output enable) low		70		85		100		120	ns
$t_{a(W)}$	Access time from write enable high		70		85		100		120	ns
$t_{v(A)}$	Output data valid after address change	20		20		20		20		ns
$t_{dis(S)}$	Output disable time after chip select (or output enable) high		50		60		60		100	ns
$t_{dis(W)}$	Output disable time after write enable low		50		60		60		100	ns

read cycle timing †

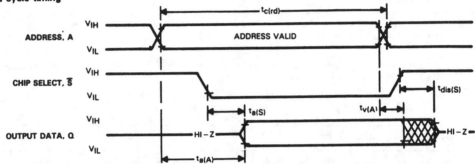

All timing reference points are 0.8 V and 2.0 V on inputs and 0.6 V and 2.2 V on outputs (90% points). Input rise and fall times equal 10 nanoseconds.

†Write enable is high for a read cycle.

early write cycle timing

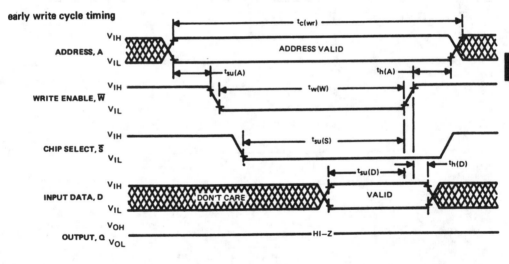

Static RAM and Memory Support Devices

8

read-write cycle timing

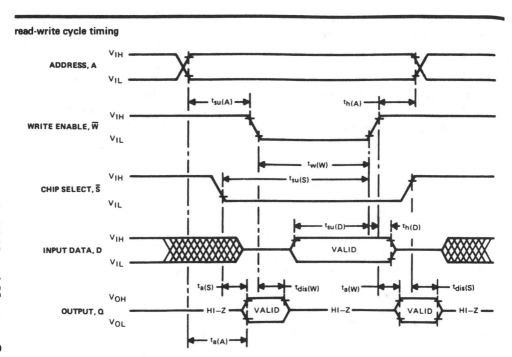

TYPICAL APPLICATION DATA

Early write cycle avoids DQ conflicts by controlling the write time with $\overline{S}$. On the diagram above, the write operation will be controlled by the leading edge of $\overline{S}$, not $\overline{W}$. Data can only be written when both $\overline{S}$ and $\overline{W}$ are low. Either $\overline{S}$ or $\overline{W}$ being high inhibits the write operation. To prevent erroneous data being written into the array, the addresses must be stable during the write cycle as defined by $t_{su(A)}$, $t_{w(W)}$, and $t_{h(A)}$.

TEXAS INSTRUMENTS
POST OFFICE BOX 225012 ● DALLAS, TEXAS 75265

- Timing from Microseconds to Hours

- Astable or Monostable Operation

- Adjustable Duty Cycle

- TTL-Compatible Output Can Sink or Source Up to 200 mA

- Functionally Interchangeable with the Signetics SE555, SE555C, SA555, NE555; Have Same Pinout

SE555C FROM TI IS NOT RECOMMENDED FOR NEW DESIGNS

description

These devices are monolithic timing circuits capable of producing accurate time delays or oscillation. In the time-delay or monostable mode of operation, the timed interval is controlled by a single external resistor and capacitor network. In the astable mode of operation, the frequency and duty cycle may be independently controlled with two external resistors and a single external capacitor.

The threshold and trigger levels are normally two-thirds and one-third, respectively, of V_{CC}. These levels can be altered by use of the control voltage terminal. When the trigger input falls below the trigger level, the flip-flop is set and the output goes high. If the trigger input is above the trigger level and the threshold input is above the threshold level, the flip-flop is reset and the output is low. The reset input can override all other inputs and can be used to initiate a new timing cycle. When the reset input goes low, the flip-flop is reset and the output goes low. Whenever the output is low, a low-impedance path is provided between the discharge terminal and ground.

The output circuit is capable of sinking or sourcing current up to 200 mA. Operation is specified for supplies of 5 to 15 V. With a 5-V supply, output levels are compatible with TTL inputs.

The SE555 and SE555C are characterized for operation over the full military range of –55°C to 125°C. The SA555 is characterized for operation from –40°C to 85°C, and the NE555 is characterized for operation from 0°C to 70°C.

SE555, SE555C . . . JG PACKAGE
SA555, NE555 . . . D, JG, OR P PACKAGE
(TOP VIEW)

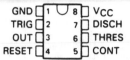

GND	1	8	V_{CC}
TRIG	2	7	DISCH
OUT	3	6	THRES
RESET	4	5	CONT

SE555, SE555C . . . FK PACKAGE
(TOP VIEW)

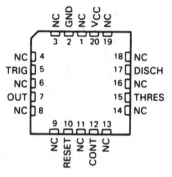

NC – No internal connection

functional block diagram

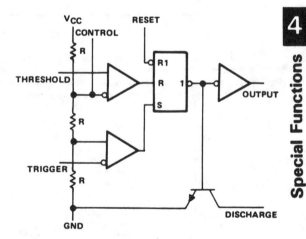

Reset can override Trigger, which can override Threshold.

4 — Special Functions

TEXAS INSTRUMENTS
POST OFFICE BOX 655012 • DALLAS, TEXAS 75265

electrical characteristics at 25 °C free-air temperature, V_{CC} = 5 V to 15 V (unless otherwise noted)

PARAMETER	TEST CONDITIONS		SE555			SE555C, SA555, NE555			UNIT
			MIN	TYP	MAX	MIN	TYP	MAX	
Threshold voltage level	V_{CC} = 15 V		9.4	10	10.6	8.8	10	11.2	V
	V_{CC} = 5 V		2.7	3.3	4	2.4	3.3	4.2	
Threshold current (see Note 2)				30	250		30	250	nA
Trigger voltage level	V_{CC} = 15 V		4.8	5	5.2	4.5	5	5.6	V
	V_{CC} = 5 V		1.45	1.67	1.9	1.1	1.67	2.2	
Trigger current	Trigger at 0 V			0.5	0.9		0.5	2	µA
Reset voltage level			0.3	0.7	1	0.3	0.7	1	V
Reset current	Reset at V_{CC}			0.1	0.4		0.1	0.4	mA
	Reset at 0 V			−0.4	−1		−0.4	−1.5	
Discharge switch off-state current				20	100		20	100	nA
Control voltage (open circuit)	V_{CC} = 15 V		9.6	10	10.4	9	10	11	V
	V_{CC} = 5 V		2.9	3.3	3.8	2.6	3.3	4	
Low-level output voltage	V_{CC} = 15 V	I_{OL} = 10 mA		0.1	0.15		0.1	0.25	V
		I_{OL} = 50 mA		0.4	0.5		0.4	0.75	
		I_{OL} = 100 mA		2	2.2		2	2.5	
		I_{OL} = 200 mA		2.5			2.5		
	V_{CC} = 5 V	I_{OL} = 5 mA		0.1	0.2		0.1	0.35	
		I_{OL} = 8 mA		0.15	0.25		0.15	0.4	
High-level output voltage	V_{CC} = 15 V	I_{OH} = −100 mA	13	13.3		12.75	13.3		V
		I_{OH} = −200 mA		12.5			12.5		
	V_{CC} = 5 V	I_{OH} = −100 mA	3	3.3		2.75	3.3		
Supply current	Output low, No load	V_{CC} = 15 V		10	12		10	15	mA
		V_{CC} = 5 V		3	5		3	6	
	Output high, No load	V_{CC} = 15 V		9	10		9	13	
		V_{CC} = 5 V		2	4		2	5	

NOTE 2: This parameter influences the maximum value of the timing resistors R_A and R_B in the circuit of Figure 12. For example, when V_{CC} = 5 V, the maximum value is R = R_A + R_B ≈ 3.4 MΩ, and for V_{CC} = 15 V, the maximum value is 10 MΩ.

operating characteristics, V_{CC} = 5 V and 15 V

PARAMETER	TEST CONDITIONS[†]		SE555			SE555C, SA555, NE555			UNIT
			MIN	TYP	MAX	MIN	TYP	MAX	
Initial error of timing interval[‡]	Each timer, monostable[§]	T_A = 25 °C		0.5	1.5		1	3	%
	Each timer, astable[¶]			1.5			2.25		
Temperature coefficient of timing interval	Each timer, monostable[§]	T_A = MIN to MAX		30	100		50		ppm/°C
	Each timer, astable[¶]			90			150		
Supply voltage sensitivity of timing interval	Each timer, monostable[§]	T_A = 25 °C		0.05	0.2		0.1	0.5	%/V
	Each timer, astable[¶]			0.15			0.3		
Output pulse rise time		C_L = 15 pF,		100	200		100	300	ns
Output pulse fall time		T_A = 25 °C		100	200		100	300	

[†] For conditions shown as MIN or MAX, use the appropriate value specified under recommended operating conditions.

[‡] Timing interval error is defined as the difference between the measured value and the average value of a random sample from each process run.

[§] Values specified are for a device in a monostable circuit similar to Figure 9, with component values as follow: R_A = 2 kΩ to 100 kΩ, C = 0.1 µF.

[¶] Values specified are for a device in an astable circuit similar to Figure 12, with component values as follow: R_A = 1 kΩ to 100 kΩ, C = 0.1 µF.

4

Special Functions

TEXAS
INSTRUMENTS
POST OFFICE BOX 655012 • DALLAS, TEXAS 75265

TYPICAL APPLICATION DATA

monostable operation

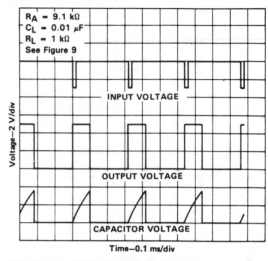

FIGURE 9. CIRCUIT FOR MONOSTABLE OPERATION

FIGURE 10. TYPICAL MONOSTABLE WAVEFORMS

For monostable operation, any of these timers may be connected as shown in Figure 9. If the output is low, application of a negative-going pulse to the trigger input sets the flip-flop ($\overline{Q}$ goes low), drives the output high, and turns off Q1. Capacitor C is then charged through R_A until the voltage across the capacitor reaches the threshold voltage of the threshold input. If the trigger input has returned to a high level, the output of the threshold comparator will reset the flip-flop ($\overline{Q}$ goes high), drive the output low, and discharge C through Q1.

Monostable operation is initiated when the trigger input voltage falls below the trigger threshold. Once initiated, the sequence ends only if the trigger input is high at the end of the timing interval. Because of the threshold level and saturation voltage of Q1, the output pulse duration is approximately $t_W = 1.1\ R_A C$. Figure 11 is a plot of the time constant for various values of R_A and C. The threshold levels and charge rates are both directly proportional to the supply voltage, V_{CC}. The timing interval is therefore independent of the supply voltage, so long as the supply voltage is constant during the time interval.

Applying a negative-going trigger pulse simultaneously to the reset and trigger terminals during the timing interval discharges C and re-initiates the cycle, commencing on the positive edge of the reset pulse. The output is held low as long as the reset pulse is low. To prevent false triggering, when the reset input is not used, it should be connected to V_{CC}.

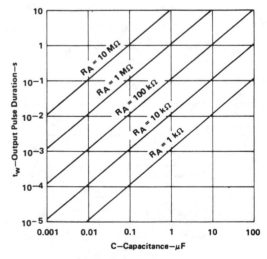

FIGURE 11. OUTPUT PULSE
DURATION vs CAPACITANCE

TEXAS
INSTRUMENTS

POST OFFICE BOX 655012 • DALLAS, TEXAS 75265

TYPICAL APPLICATION DATA

astable operation

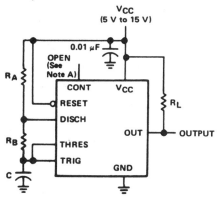

NOTE A: Decoupling the control voltage input to ground with a capacitor may improve operation. This should be evaluated for individual applications.

FIGURE 12. CIRCUIT FOR ASTABLE OPERATION

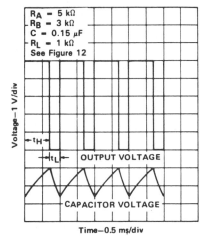

FIGURE 13. TYPICAL ASTABLE WAVEFORMS

As shown in Figure 12, adding a second resistor, R_B, to the circuit of Figure 9 and connecting the trigger input to the threshold input causes the timer to self-trigger and run as a multivibrator. The capacitor C will charge through R_A and R_B and then discharge through R_B only. The duty cycle may be controlled, therefore, by the values of R_A and R_B.

This astable connection results in capacitor C charging and discharging between the threshold-voltage level ($\approx 0.67 \cdot V_{CC}$) and the trigger-voltage level ($\approx 0.33 \cdot V_{CC}$). As in the monostable circuit, charge and discharge times (and therefore the frequency and duty cycle) are independent of the supply voltage.

Figure 13 shows typical waveforms generated during astable operation. The output high-level duration t_H and low-level duration t_L may be calculated as follows:

$$t_H = 0.693 (R_A + R_B) C$$

$$t_L = 0.693 (R_B) C$$

Other useful relationships are shown below.

$$period = t_H + t_L = 0.693 (R_A + 2R_B) C$$

$$frequency \approx \frac{1.44}{(R_A + 2R_B) C}$$

$$Output\ driver\ duty\ cycle = \frac{t_L}{t_H + t_L} = \frac{R_B}{R_A + 2R_B}$$

$$Output\ waveform\ duty\ cycle = \frac{t_H}{t_H + t_L} = 1 - \frac{R_B}{R_A + 2R_B}$$

$$Low\text{-}to\text{-}high\ ratio = \frac{t_L}{t_H} = \frac{R_B}{R_A + R_B}$$

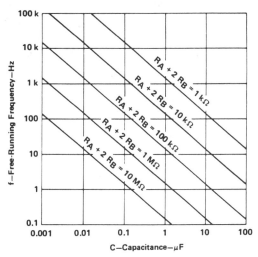

FIGURE 14. FREE-RUNNING FREQUENCY

Special Functions 4

TYPICAL APPLICATION DATA

sequential timer

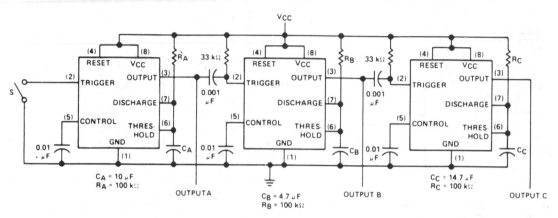

S closes momentarily at t = 0.

FIGURE 22. SEQUENTIAL TIMER CIRCUIT

Many applications, such as computers, require signals for initializing conditions during start-up. Other applications, such as test equipment, require activation of test signals in sequence. These timing circuits may be connected to provide such sequential control. The timers may be used in various combinations of astable or monostable circuit connections, with or without modulation, for extremely flexible waveform control. Figure 22 illustrates a sequencer circuit with possible applications in many systems, and Figure 23 shows the output waveforms.

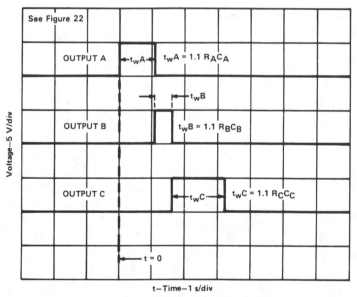

FIGURE 23. SEQUENTIAL TIMER WAVEFORMS

Special Functions 4

- **8-Bit Resolution**
- **Ratiometric Conversion**
- **100-μs Conversion Time**
- **135-ns Access Time**
- **No Zero Adjust Requirement**
- **On-Chip Clock Generator**
- **Single 5-V Power Supply**
- **Operates with Microprocessor or as Stand-Alone**
- **Designed to be Interchangeable with National Semiconductor and Signetics ADC0804**

N DUAL-IN-LINE PACKAGE
(TOP VIEW)

$\overline{CS}$	1	20	V_{CC} (OR REF)
$\overline{RD}$	2	19	CLK OUT
$\overline{WR}$	3	18	DB0 (LSB)
CLK IN	4	17	DB1
$\overline{INTR}$	5	16	DB2
IN +	6	15	DB3
IN −	7	14	DB4
ANLG GND	8	13	DB5
REF/2	9	12	DB6
DGTL GND	10	11	DB7 (MSB)

DATA OUTPUTS (DB3–DB7)

2 Data Sheets

description

The ADC0804 is a CMOS 8-bit successive-approximation analog-to-digital converter that uses a modified potentiometric (256R) ladder. The ADC0804 is designed to operate from common microprocessor control buses, with the three-state output latches driving the data bus. The ADC0804 can be made to appear to the microprocessor as a memory location or an I/O port. Detailed information on interfacing to most popular microprocessors is readily available from the factory.

A differential analog voltage input allows increased common-mode rejection and offset of the zero-input analog voltage value. Although a reference input (REF/2) is available to allow 8-bit conversion over smaller analog voltage spans or to make use of an external reference, ratiometric conversion is possible with the REF/2 input open. Without an external reference, the conversion takes place over a span from V_{CC} to analog ground (ANLG GND). The ADC0804 can operate with an external clock signal or, with an additional resistor and capacitor, can operate using an on-chip clock generator.

The ADC0804I is characterized for operation from −40 °C to 85 °C. The ADC0804C is characterized for operation from 0 °C to 70 °C.

TEXAS INSTRUMENTS
POST OFFICE BOX 655012 • DALLAS, TEXAS 75265

functional block diagram (positive logic)

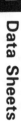 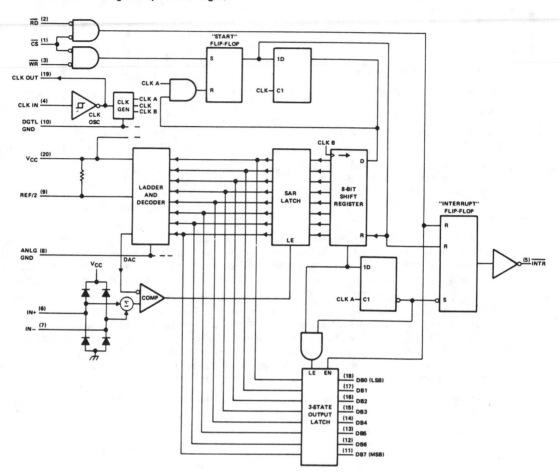

TEXAS
INSTRUMENTS
POST OFFICE BOX 655012 • DALLAS, TEXAS 75265

absolute maximum ratings over operating free-air temperature range (unless otherwise noted)

Supply voltage, V_{CC} (see Note 1) .... 6.5 V
Input voltage range: $\overline{CS}$, $\overline{RD}$, $\overline{WR}$ .... −0.3 V to 18 V
      other inputs .... −0.3 V to V_{CC} + 0.3 V
Output voltage range .... −0.3 V to V_{CC} + 0.3 V
Operating free-air temperature range: ADC0804I .... −40°C to 85°C
            ADC0804C .... 0°C to 70°C
Storage temperature range .... −65°C to 150°C
Lead temperature 1,6 mm (1/16 inch) from case for 10 seconds .... 260°C

NOTE 1: All voltage values are with respect to digital ground (DGTL GND) with DGTL GND and ANLG GND connected together (unless otherwise noted).

recommended operating conditions

		MIN	NOM	MAX	UNIT
Supply voltage, V_{CC}		4.5	5	6.3	V
Voltage at REF/2, $V_{REF/2}$ (see Note 2)		0.25	2.5		V
High-level input voltage at $\overline{CS}$, $\overline{RD}$, or $\overline{WR}$, V_{IH}		2		15	V
Low-level input voltage at $\overline{CS}$, $\overline{RD}$, or $\overline{WR}$, V_{IL}				0.8	V
Analog ground voltage (see Note 3)		−0.05	0	1	V
Analog input voltage (see Note 4)		−0.05		V_{CC}+0.05	V
Clock input frequency, f_{clock} (see Note 5)		100	640	1460	kHz
Duty cycle for $f_{clock} \geq$ 640 kHz (see Note 5)		40		60	%
Pulse duration clock input (high or low) for $f_{clock} <$ 640 kHz, $t_{w(CLK)}$ (see Note 5)		275	781		ns
Pulse duration, $\overline{WR}$ input low (start conversion), $t_{w(WR)}$		100			ns
Operating free-air temperature, T_A	ADC0804I	−40		85	°C
	ADC0804C	0		70	

NOTES: 2. The internal reference voltage is equal to the voltage applied to REF/2, or approximately equal to one-half of the V_{CC} when REF/2 is left open. The voltage at REF/2 should be one-half the full-scale differential input voltage between the analog inputs. Thus, the differential input voltage when REF/2 is open and V_{CC} = 5 V is 0 to 5 V. VREF/2 for an input voltage range from 0.5 V to 3.5 V (full-scale differential voltage of 3 V) is 1.5 V.
    3. These values are with respect to DGTL GND.
    4. When the differential input voltage ($V_{IN+} - V_{in-}$) is less than or equal to 0 V, the output code is 0000 0000.
    5. Total unadjusted error is specified only at an f_{clock} of 640 kHz with a duty cycle of 40% to 60% (pulse duration 625 ns to 937 ns). For frequencies above this limit or pulse duration below 625 ns, error may increase. The duty cycle limits should be observed for an f_{clock} greater than 640 kHz. Below 640 kHz, this duty cycle limit can be exceeded provided $t_{w(CLK)}$ remains within limits.

2

Data Sheets

TEXAS
INSTRUMENTS
POST OFFICE BOX 655012 • DALLAS, TEXAS 75265

electrical characteristics over recommended operating free-air temperature range, V_{CC} = 5 V, f_{clock} = 640 kHz, REF/2 = 2.5 V (unless otherwise noted)

PARAMETER			TEST CONDITIONS	MIN	TYP†	MAX	UNIT
V_{OH}	High-level output voltage	All outputs	V_{CC} = 4.75 V, I_{OH} = −360 μA	2.4			V
		DB and $\overline{INTR}$	V_{CC} = 4.75 V, I_{OH} = −10 μA	4.5			
V_{OL}	Low-level output voltage	Data outputs	V_{CC} = 4.75 V, I_{OL} = 1.6 mA			0.4	V
		$\overline{INTR}$ output	V_{CC} = 4.75 V, I_{OL} = 1 mA			0.4	
		CLK OUT	V_{CC} = 4.75 V, I_{OL} = 360 μA			0.4	
V_{T+}	Clock positive-going threshold voltage			2.7	3.1	3.5	V
V_{T-}	Clock negative-going threshold voltage			1.5	1.8	2.1	V
$V_{T+} - V_{T-}$	Clock input hysteresis			0.6	1.3	2	V
I_{IH}	High-level input current				0.005	1	μA
I_{IL}	Low-level input current				−0.005	−1	μA
I_{OZ}	Off-state output current		V_O = 0			−3	μA
			V_O = 5 V			3	
I_{OHS}	Short-circuit output current	Output high	V_O = 0, T_A = 25°C	−4.5	−6		mA
I_{OLS}	Short-circuit output current	Output low	V_O = 5 V, T_A = 25°C	9	16		mA
I_{CC}	Supply current plus reference current		REF/2 open, $\overline{CS}$ at 5 V, T_A = 25°C		1.9	2.5	mA
$R_{REF/2}$	Input resistance to reference ladder		See Note 6	1	1.3		kΩ
C_i	Input capacitance (control)				5	7.5	pF
C_o	Output capacitance (DB)				5	7.5	pF

operating characteristics over recommended operating free-air temperature range, V_{CC} = 5 V, $V_{REF/2}$ = 2.5 V, f_{clock} = 640 kHz (unless otherwise noted)

PARAMETER		TEST CONDITIONS	MIN	TYP†	MAX	UNIT
	Supply-voltage-variation error (See Notes 2 and 7)	V_{CC} = 4.5 V to 5.5 V		±1/16	±1/8	LSB
	Total unadjusted error (See Notes 7 and 8)	$V_{REF/2}$ = 2.5 V			±1	LSB
	DC common-mode error (See Note 8)			±1/16	±1/8	LSB
t_{en}	Output enable time	C_L = 100 pF		135	200	ns
t_{dis}	Output disable time	C_L = 10 pF, R_L = 10 kΩ		125	200	ns
$t_{d(INTR)}$	Delay time to reset $\overline{INTR}$			300	450	ns
t_{conv}	Conversion cycle time (See Note 9)	f_{clock} = 100 kHz to 1.46 MHz	65½		72½	clock cycles
	Conversion time		103		114	μs
CR	Free-running conversion rate	$\overline{INTR}$ connected to $\overline{WR}$, $\overline{CS}$ at 0 V			8827	conv/s

†All typical values are at T_A = 25°C.

NOTES: 2. The internal reference voltage is equal to the voltage applied to REF/2, or approximately equal to one-half of the V_{CC} when REF/2 is left open. The voltage at REF/2 should be one-half the full-scale differential input voltage between the analog inputs. Thus, the differential input voltage when REF/2 is open and V_{CC} = 5 V is 0 to 5 V. $V_{REF/2}$ for an input voltage range from 0.5 V to 3.5 V (full-scale differential voltage of 3 V) is 1.5 V.
     6. The resistance is calculated from the current drawn from a 5-V supply applied to pins 8 and 9.
     7. These parameters are specified for the recommended analog input voltage range.
     8. All errors are measured with reference to an ideal straight line through the end-points of the analog-to-digital transfer characteristic.
     9. Although internal conversion is completed in 64 clock periods, a $\overline{CS}$ or $\overline{WR}$ low-to-high transition is followed by 1 to 8 clock periods before conversion starts. After conversion is completed, part of another clock period is required before a high-to-low transition of $\overline{INTR}$ completes the cycle.

timing diagrams

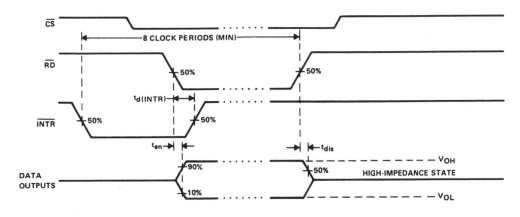

READ OPERATION TIMING DIAGRAM

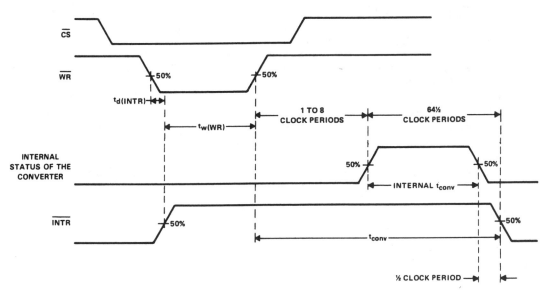

WRITE OPERATION TIMING DIAGRAM

PRINCIPLES OF OPERATION

The ADC0804 contains a circuit equivalent to a 256-resistor network. Analog switches are sequenced by successive approximation logic to match an analog differential input voltage ($V_{in+} - V_{in-}$) to a corresponding tap on the 256-resistor network. The most-significant bit (MSB) is tested first. After eight comparisons (64 clock periods), an 8-bit binary code (1111 1111 = full scale) is transferred to an output latch and the interrupt ($\overline{INTR}$) output goes low. The device can be operated in a free-running mode by connecting the $\overline{INTR}$ output to the write ($\overline{WR}$) input and holding the conversion start ($\overline{CS}$) input at a low level. To ensure start-up under all conditions, a low-level $\overline{WR}$ input is required during the power-up cycle. Taking $\overline{CS}$ low anytime after that will interrupt a conversion in process.

When the $\overline{WR}$ input goes low, the ADC0804 successive approximation register (SAR) and 8-bit shift register are reset. As long as both $\overline{CS}$ and $\overline{WR}$ remain low, the ADC0804 remains in a reset state. One to eight clock periods after $\overline{CS}$ or $\overline{WR}$ makes a low-to-high transition, conversion starts.

When the $\overline{CS}$ and $\overline{WR}$ inputs are low, the start flip-flop is set and the interrupt flip-flop and 8-bit register are reset. The next clock pulse transfers a logic high to the output of the start flip-flop. The logic high is ANDed with the next clock pulse, placing a logic high on the reset input of the start flip-flop. If either $\overline{CS}$ or $\overline{WR}$ have gone high, the set signal to the start flip-flop is removed, causing it to be reset. A logic high is placed on the D input of the 8-bit shift register and the conversion process is started. If the $\overline{CS}$ and $\overline{WR}$ inputs are still low, the start flip-flop, the 8-bit shift register, and the SAR remain reset. This action allows for wide $\overline{CS}$ and $\overline{WR}$ inputs with conversion starting from one to eight clock periods after one of the inputs goes high.

When the logic high input has been clocked through the 8-bit shift register, completing the SAR search, it is applied to an AND gate controlling the output latches and to the D input of a flip-flop. On the next clock pulse, the digital word is transferred to the three-state output latches and the interrupt flip-flop is set. The output of the interrupt flip-flop is inverted to provide an $\overline{INTR}$ output that is high during conversion and low when the conversion is completed.

When a low is at both the $\overline{CS}$ and $\overline{RD}$ inputs, an output is applied to the DB0 through DB7 outputs and the interrupt flip-flop is reset. When either the $\overline{CS}$ or $\overline{RD}$ inputs return to a high state, the DB0 through DB7 outputs are disabled (returned to the high-impedance state). The interrupt flip-flop remains reset.

TEXAS
INSTRUMENTS
POST OFFICE BOX 655012 • DALLAS, TEXAS 75265

DACPORT
Low-Cost Complete
μP-Compatible 8-Bit DAC

AD557

FEATURES
Complete 8-Bit DAC
Voltage Output – 0 to 2.56V
Internal Precision Band-Gap Reference
Single-Supply Operation: +5V (±10%)
Full Microprocessor Interface
Fast: 1μs Voltage Settling to ±1/2LSB
Low Power: 75mW
No User Trims Required
Guaranteed Monotonic Over Temperature
All Errors Specified T_{min} to T_{max}
Small 16-Pin DIP or 20-Pin PLCC Package
Low Cost

AD557 FUNCTIONAL BLOCK DIAGRAM

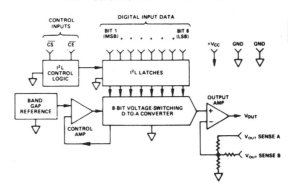

PRODUCT DESCRIPTION
The AD557 DACPORT™ is a complete voltage-output 8-bit digital-to-analog converter, including output amplifier, full microprocessor interface and precision voltage reference on a single monolithic chip. No external components or trims are required to interface, with full accuracy, an 8-bit data bus to an analog system.

The low cost and versatility of the AD557 DACPORT are the result of continued development in monolithic bipolar technologies.

The complete microprocessor interface and control logic is implemented with integrated injection logic (I²L), an extremely dense and low-power logic structure that is process-compatible with linear bipolar fabrication. The internal precision voltage reference is the patented low-voltage band-gap circuit which permits full-accuracy performance on a single +5V power supply. Thin-film silicon-chromium resistors provide the stability required for guaranteed monotonic operation over the entire operating temperature range, while laser-wafer trimming of these thin-film resistors permits absolute calibration at the factory to within ±2.5LSB; thus, no user-trims for gain or offset are required. A new circuit design provides voltage settling to ±1/2LSB for a full-scale step in 800ns.

The AD557 is available in two package configurations. The AD557JN is packaged in a 16-pin plastic, 0.3"-wide DIP. For surface mount applications, the AD557JP is packaged in a 20-pin JEDEC standard PLCC. Both versions are specified over the operating temperature range of 0 to +70°C.

DACPORT is a trademark of Analog Devices, Inc.
Covered by U.S. Patent Nos. 3,887,863; 3,685,045; 4,323,795; other patents pending.

PRODUCT HIGHLIGHTS
1. The 8-bit I²L input register and fully microprocessor-compatible control logic allow the AD557 to be directly connected to 8- or 16-bit data buses and operated with standard control signals. The latch may be disabled for direct DAC interfacing.

2. The laser-trimmed on-chip SiCr thin-film resistors are calibrated for absolute accuracy and linearity at the factory. Therefore, no user trims are necessary for full rated accuracy over the operating temperature range.

3. The inclusion of a precision low-voltage band-gap reference eliminates the need to specify and apply a separate reference source.

4. The AD557 is designed and specified to operate from a single +4.5V to +5.5V power supply.

5. Low digital input currents, 100μA max, minimize bus loading. Input thresholds are TTL/low voltage CMOS compatible.

6. The single-chip, low power I²L design of the AD557 is inherently more reliable than hybrid multichip or conventional single-chip bipolar designs.

SPECIFICATIONS (@ T_A = +25°C, V_{CC} = +5V unless otherwise specified)

Model	AD557J Min	Typ	Max	Units
RESOLUTION			8	Bits
RELATIVE ACCURACY[1]				
0 to +70°C		±1/2	1	LSB
OUTPUT				
Ranges		0 to +2.56		V
Current Source	+5			mA
Sink		Internal Passive Pull-Down to Ground[2]		
OUTPUT SETTLING TIME[3]		0.8	1.5	µs
FULL SCALE ACCURACY[4]				
@25°C		±1.5	±2.5	LSB
T_{min} to T_{max}		±2.5	±4.0	LSB
ZERO ERROR				
@25°C			±1	LSB
T_{min} to T_{max}			±3	LSB
MONOTONICITY[5]				
T_{min} to T_{max}		Guaranteed		
DIGITAL INPUTS				
T_{min} to T_{max}				
Input Current			±100	µA
Data Inputs, Voltage				
Bit On – Logic "1"	2.0			V
Bit On – Logic "0"	0		0.8	V
Control Inputs, Voltage				
On – Logic "1"	2.0			V
On – Logic "0"	0		0.8	V
Input Capacitance		4		pF
TIMING[6]				
t_W Strobe Pulse Width	225			ns
T_{min} to T_{max}	**300**			ns
t_{DH} Data Hold Time	10			ns
T_{min} to T_{max}	**10**			ns
t_{DS} Data Setup Time	225			ns
T_{min} to T_{max}	**300**			ns
POWER SUPPLY				
Operating Voltage Range (V_{CC})				
2.56 Volt Range	+4.5		+5.5	V
Current (I_{CC})		15	25	mA
Rejection Ratio			**0.03**	%/%
POWER DISSIPATION, V_{CC} = 5V		75	125	mW
OPERATING TEMPERATURE RANGE	0		+70	°C

NOTES
[1]Relative Accuracy is defined as the deviation of the code transition points from the ideal transfer point on a straight line from the offset to the full scale of the device. See "Measuring Offset Error" on AD558 data sheet.
[2]Passive pull-down resistance is 2kΩ.
[3]Settling time is specified for a positive-going full-scale step to ±1/2LSB. Negative-going steps to zero are slower, but can be improved with an external pull-down.
[4]The full-scale output voltage is 2.55V and is guaranteed with a +5V supply.
[5]A monotonic converter has a maximum differential linearity error of ±1LSB.
[6]See Figure 7.
Specifications shown in **boldface** are tested on all production units at final electrical test.
Specifications subject to change without notice.

ABSOLUTE MAXIMUM RATINGS*
V_{CC} to Ground	0V to +18V
Digital Inputs (Pins 1-10)	0 to +7.0V
V_{OUT}	Indefinite Short to Ground, Momentary Short to V_{CC}
Power Dissipation	450mW
Storage Temperature Range	
N/P (Plastic) Packages	−25°C to +100°C
Lead Temperature (soldering, 10 sec)	300°C

PIN CONFIGURATIONS

DIP

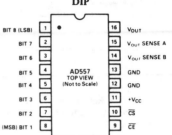

PLCC

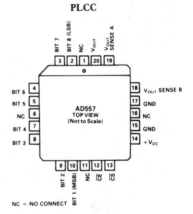

NC = NO CONNECT

AD557 ORDERING GUIDE

Model	Package Options*	Temperature
AD557JN	Plastic (N-16)	0 to +70°C
AD557JP	PLCC (P-20A)	0 to +70°C

*See Section 14 for package outline information.

Thermal Resistance
Junction to Ambient/Junction to Case
N/P (Plastic) Packages . . . . 140/55°C/W

*Stresses above those listed under "Absolute Maximum Ratings" may cause permanent damage to the device. This is a stress rating only and functional operation of the device at these or any other conditions above those indicated in the operational sections of this specification is not implied. Exposure to absolute maximum rating conditions for extended periods may affect device reliability.

CIRCUIT DESCRIPTION

The AD557 consists of four major functional blocks fabricated on a single monolithic chip (see Figure 1). The main D/A converter section uses eight equally weighted laser-trimmed current sources switched into a silicon-chromium thin-film R/2R resistor ladder network to give a direct but unbuffered 0mV to 400mV output range. The transistors that form the DAC switches are PNPs; this allows direct positive-voltage logic interface and a zero-based output range.

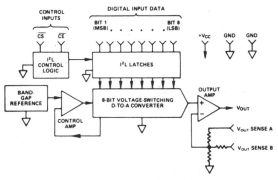

Figure 1. Functional Block Diagram

The high-speed output buffer amplifier is operated in the noninverting mode with gain determined by the user-connections at the output range select pin. The gain-setting application resistors are thin film laser trimmed to match and track the DAC resistors and to assure precise initial calibration of the output range, 0V to 2.56V. The amplifier output stage is an NPN transistor with passive pull-down for zero-based output capability with a single power supply.

The internal precision voltage reference is of the patented band-gap type. This design produces a reference voltage of 1.2V and thus, unlike 6.3V temperature-compensated zeners, may be operated from a single, low-voltage logic power supply. The microprocessor interface logic consists of an 8-bit data latch and control circuitry. Low power, small geometry and high speed are advantages of the I^2L design as applied to this section. I^2L is bipolar process compatible so that the performance of the analog sections need not be compromised to provide on-chip logic capabilities. The control logic allows the latches to be operated from a decoded microprocessor address and write signal. If the application does not involve a μP or data bus, wiring $\overline{CS}$ and $\overline{CE}$ to ground renders the latches "transparent" for direct DAC access.

Digital Input Code			Output
Binary	Hexadecimal	Decimal	Voltage
0000 0000	00	0	0
0000 0001	01	1	0.010V
0000 0010	02	2	0.020V
0000 1111	0F	15	0.150V
0001 0000	10	16	0.160V
0111 1111	7F	127	1.270V
1000 0000	80	128	1.280V
1100 0000	C0	192	1.920V
1111 1111	FF	255	2.55V

CONNECTING THE AD557

The AD557 has been configured for low cost and ease of application. All reference, output amplifier and logic connections are made internally. In addition, all calibration trims are performed at the factory assuring specified accuracy without user trims. The only connection decision to be made by the user is whether the output range desired is unipolar or bipolar. Clean circuit board layout is facilitated by isolating all digital bit inputs on one side of the package; analog outputs are on the opposite side.

UNIPOLAR 0 TO +2.56V OUTPUT RANGE

Figure 2 shows the configuration for the 0 to +2.56V full-scale output range. Because of its precise factory calibration, the AD557 is intended to be operated without user trims for gain and offset; therefore, no provisions have been made for such user trims. If a small increase in scale is required, however, it may be accomplished by slightly altering the effective gain of the output buffer. A resistor in series with V_{OUT} SENSE will increase the output range. Note that decreasing the scale by putting a resistor in series with GND will not work properly due to the code-dependent currents in GND. Adjusting offset by injecting dc at GND is not recommended for the same reason.

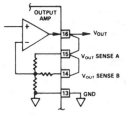

Figure 2. 0 to 2.56V Output Range

BIPOLAR −1.28V TO +1.28V OUTPUT RANGE

The AD557 was designed for operation from a single power supply and is thus capable of providing only a unipolar 0 to +2.56V output range. If a negative supply is available, bipolar output ranges may be achieved by suitable output offsetting and scaling. Figure 3 shows how a ±1.28V output range may be achieved when a −5V power supply is available. The offset is provided by the AD589 precision 1.2V reference which will operate from a +5V supply. The AD711 output amplifier can provide the necessary ±1.28V output swing from ±5V supplies. Coding is complementary offset binary.

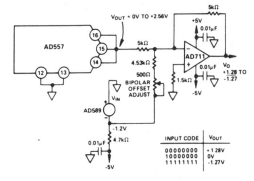

Figure 3. Bipolar Operation of AD557 from ±5V Supplies

Applications

GROUNDING AND BYPASSING

All precision converter products require careful application of good grounding practices to maintain full rated performance. Because the AD557 is intended for application in microcomputer systems where digital noise is prevalent, special care must be taken to assure that its inherent precision is realized.

The AD557 has two ground (common) pins; this minimizes ground drops and noise in the analog signal path. Figure 4 shows how the ground connections should be made.

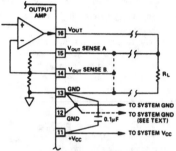

Figure 4. Recommended Grounding and Bypassing

It is often advisable to maintain separate analog and digital grounds throughout a complete system, tying them common in one place only. If the common tie-point is remote and accidental disconnection of that one common tie-point occurs due to card removal with power on, a large differential voltage between the two commons could develop. To protect devices that interface to both digital and analog parts of the system, such as the AD557, it is recommended that common ground tie-points should be provided at *each* such device. If only one system ground can be connected directly to the AD557, it is recommended that analog common be selected.

USING A "FALSE" GROUND

Many applications, such as disk drives, require servo control voltages that swing on either side of a "false" ground. This ground is usually created by dividing the + 12V supply equally and calling the midpoint voltage "ground."

Figure 5 shows an easy and inexpensive way to implement this. The AD586 is used to provide a stable 5V reference from the system's + 12V supply. The op amp shown likewise operates from a single (+ 12V) supply available in the system. The resulting output at the V_{OUT} node is ± 2.5V around the "false" ground point of 5V. AD557 input code vs. V_{OUT} is shown in Figure 6.

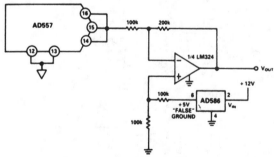

Figure 5. Level Shifting the AD557 Output Around a "False" Ground

TIMING AND CONTROL

The AD557 has data input latches that simplify interface to 8- and 16-bit data buses. These latches are controlled by Chip Enable ($\overline{CE}$) and Chip Select ($\overline{CS}$) inputs. $\overline{CE}$ and $\overline{CS}$ are internally "NORed" so that the latches transmit input data to the DAC section when both $\overline{CE}$ and $\overline{CS}$ are at Logic "0". If the application does not involve a data bus, a "00" condition allows for direct operation of the DAC. When either $\overline{CE}$ or $\overline{CS}$ go to Logic "1," the input data is latched into the registers and held until both $\overline{CE}$ and $\overline{CS}$ return to "0." (Unused $\overline{CE}$ or $\overline{CS}$ inputs should be tied to ground.) The truth table is given in Table I. The logic function is also shown in Figure 6.

Input Data	$\overline{CE}$	$\overline{CS}$	DAC Data	Latch Condition
0	0	0	0	"transparent"
1	0	0	1	"transparent"
0	∫	0	0	latching
1	∫	0	1	latching
0	0	∫	0	latching
1	0	∫	1	latching
X	1	X	previous data	latched
X	X	1	previous data	latched

Notes: X = Does not matter
∫ = Logic Threshold at Positive-Going Transition

Table I. AD557 Control Logic Truth Table

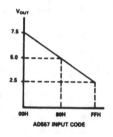

Figure 6. AD557 Input Code vs. Level Shifted Output in "False" Ground Configuration

In a level-triggered latch such as that used in the AD557, there is an interaction between the data setup and hold times and the width of the enable pulse. In an effort to reduce the time required to test all possible combinations in production, the AD557 is tested with $T_{DS} = T_W = 225ns$ at 25°C and 300ns at T_{min} and T_{max}, with $T_{DH} = 10ns$ at all temperatures. Failure to comply with these specifications may result in data not being latched properly.

Figure 7 shows the timing for the data and control signals, $\overline{CE}$ and $\overline{CS}$ are identical in timing as well as in function.

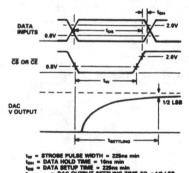

t_{w} = STROBE PULSE WIDTH = 225ns min
t_{DH} = DATA HOLD TIME = 10ns min
t_{DS} = DATA SETUP TIME = 225ns min
$t_{SETTLING}$ = DAC OUTPUT SETTLING TIME TO ± 1/2 LSB

Figure 7. AD557 Timing

GAL®16V8A
GAL®20V8A

Generic Array Logic™
U.S. Patents 4,761,768 and 4,766,569

FEATURES

- **HIGH PERFORMANCE E²CMOS™ TECHNOLOGY**
  - 10 ns Maximum Propagation Delay
  - Fmax = 62.5 MHz
  - 8 ns Maximum from Clock Input to Data Output
  - TTL Compatible 24 mA Outputs
  - UltraMOS® III Advanced CMOS Technology

- **50% REDUCTION IN POWER**
  - 75mA Typ I_{cc}

- **E² CELL TECHNOLOGY**
  - Reconfigurable Logic
  - Reprogrammable Cells
  - 100% Tested/Guaranteed 100% Yields
  - High Speed Electrical Erasure (<50ms)
  - 20 Year Data Retention

- **EIGHT OUTPUT LOGIC MACROCELLS**
  - Maximum Flexibility for Complex Logic Designs
  - Programmable Output Polarity
  - GAL16V8A Emulates 20-pin PAL® Devices with Full Function/Fuse Map/Parametric Compatibility
  - GAL20V8A Emulates 24-pin PAL® Devices with Full Function/Fuse Map/Parametric Compatibility

- **PRELOAD AND POWER-ON RESET OF ALL REGISTERS**
  - 100% Functional Testability

- **ELECTRONIC SIGNATURE FOR IDENTIFICATION**

DESCRIPTION

The GAL16V8A and GAL20V8A, at 10 ns maximum propagation delay time, combine a high performance CMOS process with Electrically Erasable (E²) floating gate technology to provide the highest speed performance available in the PLD market. CMOS circuitry allows the GAL16V8A and GAL20V8A to consume just 75mA typical I_{cc} which represents a 50% savings in power when compared to their bipolar counterparts. The E² technology offers high speed (50ms) erase times, providing the ability to reprogram or reconfigure the devices quickly and efficiently.

The generic architecture provides maximum design flexibility by allowing the Output Logic Macrocell (OLMC) to be configured by the user. The GAL16V8A and GAL20V8A are capable of emulating standard 20 and 24-pin PAL® devices. The GAL16V8A is capable of emulating standard 20-pin PAL architectures with full function/fuse map/parametric compatibility. The GAL20V8A is capable of emulating standard 24-pin PAL architectures with full function/fuse map/parametric compatibility. On the right is a table listing the PAL architectures that the GAL16V8A and GAL20V8A can replace.

Unique test circuitry and reprogrammable cells allow complete AC, DC, and functional testing during manufacture. Therefore, Lattice guarantees 100% field programmability and functionality of all GAL products. Lattice also guarantees 100 erase/rewrite cycles and that data retention exceeds 20 years.

GAL16V8A / GAL20V8A BLOCK DIAGRAM

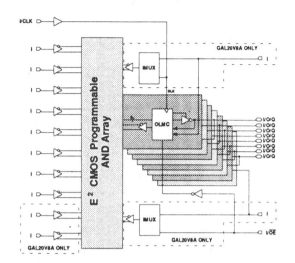

GAL16V8A / GAL20V8A ARCHITECTURE EMULATION

GAL20V8A PAL Architecture Emulation	GAL16V8A PAL Architecture Emulation
20L8	16L8
20H8	16H8
20R8	16R8
20R6	16R6
20R4	16R4
20P8	16P8
20RP8	16RP8
20RP6	16RP6
20RP4	16RP4
14L8	10L8
16L6	12L6
18L4	14L4
20L2	16L2
14H8	10H8
16H6	12H6
18H4	14H4
20H2	16H2
14P8	10P8
16P6	12P6
18P4	14P4
20P2	16P2

LATTICE SEMICONDUCTOR CORP., PO BOX 2500, PORTLAND, OREGON 97208-2500, U.S.A.
Tel. (503) 681-0118; 1-800-FASTGAL; FAX (503) 681-3037

May 1989

GAL16V8A BLOCK DIAGRAM

GAL20V8A BLOCK DIAGRAM

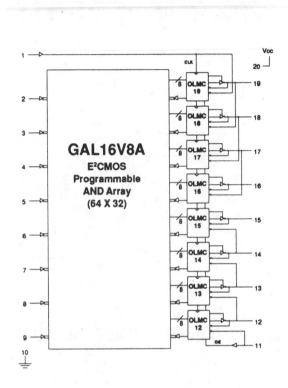

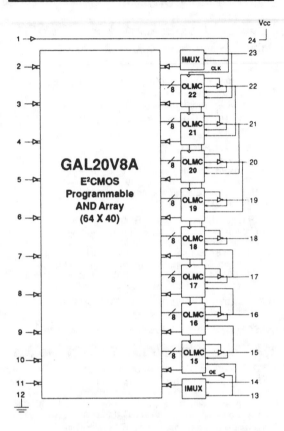

GAL16V8A PIN CONFIGURATION

GAL20V8A PIN CONFIGURATION

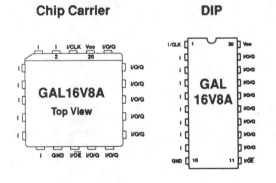

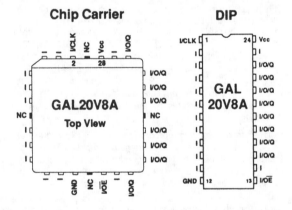

GAL16V8A LOGIC DIAGRAM

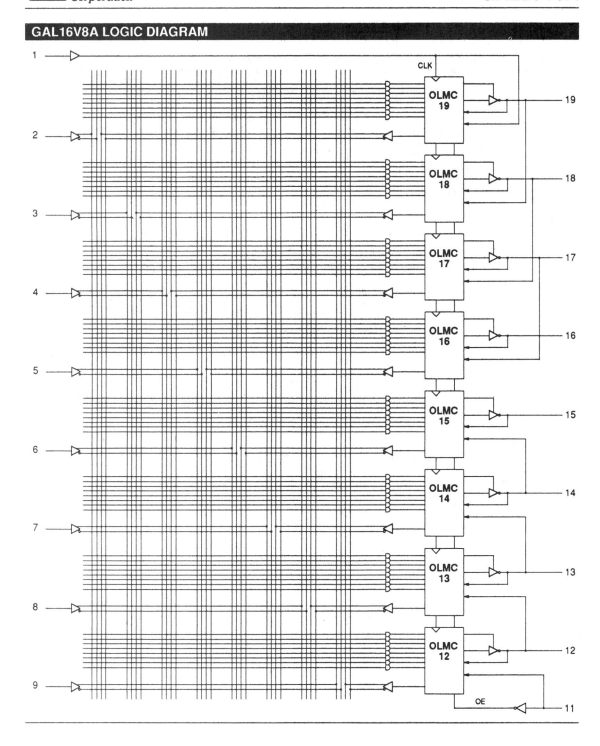

OUTPUT LOGIC MACROCELL (OLMC)

The following discussion pertains to configuring the output logic macrocell. It should be noted that actual implementation is accomplished by development software/hardware and is completely transparent to the user.

There are three OLMC configuration modes possible: registered, complex, and simple. These are illustrated in the diagrams on the following pages. You cannot mix modes, either all OLMCs are simple, complex, or registered (in registered mode the output can be combinational or registered).

The outputs of the AND array are fed into an OLMC, where each output can be individually set to active high or active low, with either combinational (asynchronous) or registered (synchronous)

configurations. A common output enable is connected to all registered outputs; or a product term can be used to provide individual output enable control for combinational outputs in the registered mode or combinational outputs in the complex mode. There is no output enable control in the small mode. The output logic macrocell provides the designer with maximum output flexibility in matching signal requirements, thus providing more functionality than possible with existing 20 and 24-pin PAL® devices.

The six valid macrocell configurations, two configurations per mode, are shown in each of the macrocell equivalent diagrams. Pin and macrocell functions are detailed in the following diagrams.

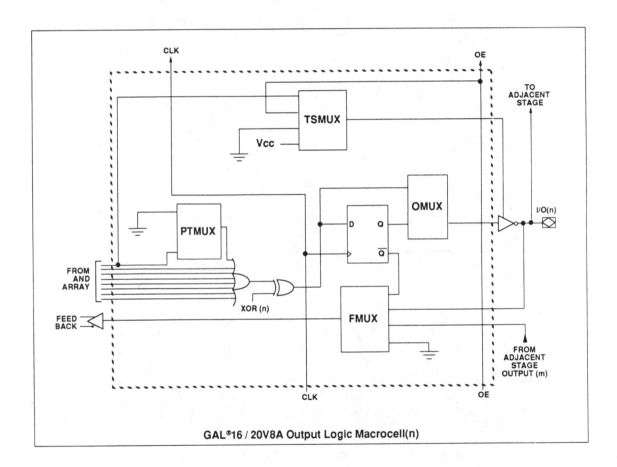

GAL®16 / 20V8A Output Logic Macrocell(n)

REGISTERED MODE

In the Registered architecture mode macrocells are configured as dedicated, registered outputs or as I/O functions.

Architecture configurations available in this mode are similar to the common 16R8, 20R6 and 16RP4 devices with various permutations of polarity, I/O and register placement.

All registered macrocells share common clock and $\overline{OE}$ control pins. Any macrocell can be configured as registered or I/O. Up to 8 registers or up to 8 I/O's are possible in this mode. Dedicated input or output functions can be implemented as sub-sets of the I/O function.

Registered outputs have 8 data product terms per output. I/O's have 7 data product terms per output.

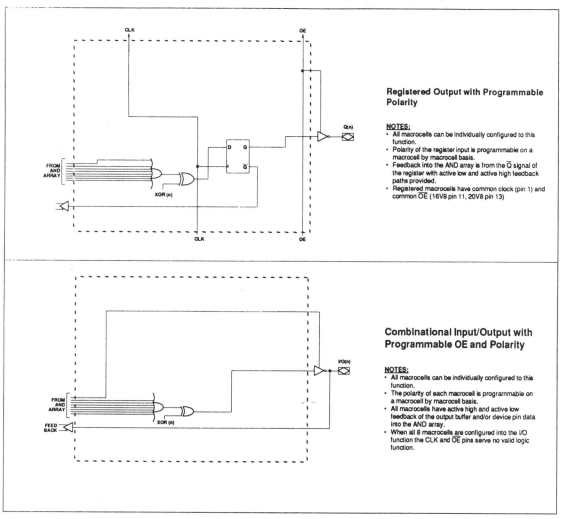

Registered Output with Programmable Polarity

NOTES:
- All macrocells can be individually configured to this function.
- Polarity of the register input is programmable on a macrocell by macrocell basis.
- Feedback into the AND array is from the $\overline{Q}$ signal of the register with active low and active high feedback paths provided.
- Registered macrocells have common clock (pin 1) and common $\overline{OE}$ (16V8 pin 11, 20V8 pin 13)

Combinational Input/Output with Programmable OE and Polarity

NOTES:
- All macrocells can be individually configured to this function.
- The polarity of each macrocell is programmable on a macrocell by macrocell basis.
- All macrocells have active high and active low feedback of the output buffer and/or device pin data into the AND array.
- When all 8 macrocells are configured into the I/O function the CLK and $\overline{OE}$ pins serve no valid logic function.

Note: The development software configures all of the architecture control bits and checks for proper pin usage automatically.

COMPLEX MODE

In the Complex architecture mode macrocells are configured as output only or I/O functions.

Architecture configurations available in this mode are similar to the common 16L8, 20L8 and 16P8 devices with programmable polarity in each macrocell.

Up to 6 I/O's are possible in this mode. Dedicated inputs or out-puts can be implemented as sub-sets of the I/O function. The two "outboard" macrocells do not have input capability. Designs requiring 8 I/O's can be implemented in the Registered mode.

All macrocells have 7 data product terms per output. One product term is used for programmable OE control. Pins 1 and 11 on a GAL16V8, and pins 1 and 13 on a GAL20V8, are always available as data inputs into the AND array.

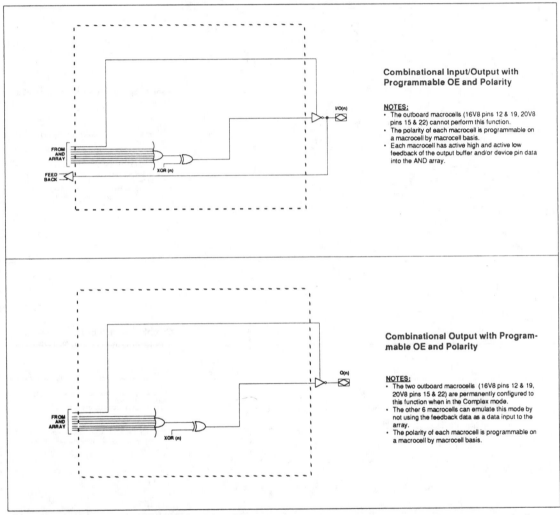

Combinational Input/Output with Programmable OE and Polarity

NOTES:
- The outboard macrocells (16V8 pins 12 & 19, 20V8 pins 15 & 22) cannot perform this function.
- The polarity of each macrocell is programmable on a macrocell by macrocell basis.
- Each macrocell has active high and active low feedback of the output buffer and/or device pin data into the AND array.

Combinational Output with Programmable OE and Polarity

NOTES:
- The two outboard macrocells (16V8 pins 12 & 19, 20V8 pins 15 & 22) are permanently configured to this function when in the Complex mode.
- The other 6 macrocells can emulate this mode by not using the feedback data as a data input to the array.
- The polarity of each macrocell is programmable on a macrocell by macrocell basis.

Note: The development software configures all of the architecture control bits and checks for proper pin usage automatically.

SIMPLE MODE

In the Simple architecture mode pins are configured as dedicated inputs or as dedicated, always active, combinational outputs.

Architecture configurations available in this mode are similar to the common 10L8, 18H4 and 16P6 devices with many permutations of generic polarity output or input choices.

All ouputs are associated with 8 data product terms. In addition, each output has programmable polarity.

Pins 1 and 11 on a GAL16V8, and pins 1 and 13 on a GAL20V8, are always available as data inputs into the AND array. The "center" two macrocells (GAL16V8 pins 15 & 16, GAL20V8 pins 18 & 19) cannot be used in the input configuration.

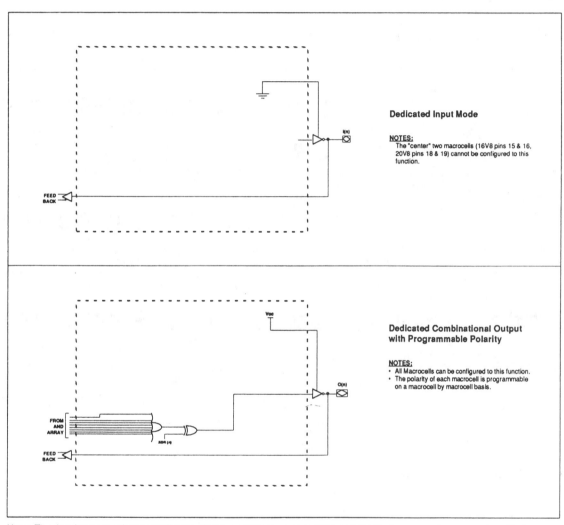

Dedicated Input Mode

NOTES:
The "center" two macrocells (16V8 pins 15 & 16, 20V8 pins 18 & 19) cannot be configured to this function.

Dedicated Combinational Output with Programmable Polarity

NOTES:
- All Macrocells can be configured to this function.
- The polarity of each macrocell is programmable on a macrocell by macrocell basis.

Note: The development software configures all of the architecture control bits and checks for proper pin usage automatically.

ELECTRONIC SIGNATURE

An electronic signature (ES) is provided with every GAL16V8A and GAL20V8A device. It contains 64 bits of reprogrammable memory that can contain user defined data. Some uses include user ID codes, revision numbers, or inventory control. The signature data is always available to the user independent of the state of the security cell.

NOTE: The ES is included in checksum calculations. Changing the ES will alter the checksum.

SECURITY CELL

A security cell is provided with every GAL16V8A and GAL20V8A device as a deterrent to unauthorized copying of the array patterns. Once programmed, this cell prevents further read access to the AND array. This cell can be erased only during a bulk erase cycle, so the original configuration can never be examined once this cell is programmed. The Electronic Signature is always available to the user, regardless of the state of this control cell.

INPUT BUFFERS

GAL16V8A and GAL20V8A devices are designed with TTL level compatible input buffers. These buffers, with their characteristically high impedance, load driving logic much less than traditional bipolar devices. This allows for a greater fan out from the driving logic.

GAL16V8A and GAL20V8A devices do not possess active pull-ups within their input structures. As a result, Lattice recommends that all unused inputs and tri-stated I/O pins be connected to another active input, V_{CC}, or GND. Doing this will tend to improve noise immunity and reduce I_{CC} for the device.

OUTPUT REGISTER PRELOAD

When testing state machine designs, all possible states and state transitions must be verified in the design, not just those required in the normal machine operations. This is because in system operation, certain events occur that may throw the logic into an illegal state (power-up, line voltage glitches, brown-outs, etc.). To test a design for proper treatment of these conditions, a way must be provided to break the feedback paths, and force any desired (i.e., illegal) state into the registers. Then the machine can be sequenced and the outputs tested for correct next state conditions.

GAL16V8A and GAL20V8A devices include circuitry that allows each registered output to be synchronously set either high or low. Thus, any present state condition can be forced for test sequencing. If necessary, approved GAL programmers capable of executing test vectors perform output register preload automatically.

LATCH-UP PROTECTION

GAL16V8A and GAL20V8A devices are designed with an on-board charge pump to negatively bias the substrate. The negative bias is of sufficient magnitude to prevent input undershoots from causing the circuitry to latch. Additionally, outputs are designed with n-channel pull-up instead of the traditional p-channel pullups to eliminate any possibility of SCR induced latching.

BULK ERASE MODE

Before writing a new pattern into a previously programmed part, the old pattern must first be erased. This erasure is done automatically by the programming hardware as part of the programming cycle and takes only 50 milliseconds.

POWER-UP RESET

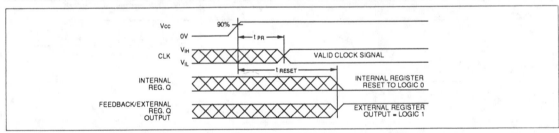

Circuitry within the GAL16V8A and GAL20V8A provides a reset signal to all registers during power-up. All internal registers will have their Q outputs set low after a specified time (t_{RESET}, 45µs MAX). As a result, the state on the registered output pins (if they are enabled through $\overline{OE}$) will always be high on power-up, regardless of the programmed polarity of the output pins. This feature can greatly simplify state machine design by providing a known state on power-up.

The timing diagram for power-up is shown above. Because of asynchronous nature of system power-up, some conditions must be met to guarantee a valid power-up reset of the GAL16V8A and GAL20V8A. First, the V_{CC} rise must be monotonic. Second, the clock input must become a proper TTL level within the specified time (t_{PR}, 100ns MAX). The registers will reset within a maximum of t_{RESET} time. As in normal system operation, avoid clocking the device until all input and feedback path setup times have been met.

ABSOLUTE MAXIMUM RATINGS(1)

Supply voltage V_{CC} .... −.5 to +7V
Input voltage applied .... −2.5 to V_{CC} +1.0V
Off-state output voltage applied .... −2.5 to V_{CC} +1.0V
Storage Temperature .... −65 to 125°C

1. Stresses above those listed under the "Absolute Maximum Ratings" may cause permanent damage to the device. These are stress only ratings and functional operation of the device at these or at any other conditions above those indicated in the operational sections of this specification is not implied (while programming, follow the programming specifications).

SWITCHING TEST CONDITIONS

Input Pulse Levels	GND to 3.0V
Input Rise and Fall Times	3ns 10% – 90%
Input Timing Reference Levels	1.5V
Output Timing Reference Levels	1.5V
Output Load	See Figure

Tri-state levels are measured 0.5V from steady-state active level.

COMMERCIAL DEVICES
Refer to AC Test Conditions:
R_2 = 390Ω
   1) R_1 = 200Ω and C_L = 50pF
   2) Active High R_1 = ∞; Active Low R_1 = 200Ω C_L = 50pF
   3) Active High R_1 = ∞; Active Low R_1 = 200Ω C_L = 5pF

MILITARY DEVICES
Refer to AC Test Conditions:
R_2 = 750Ω
   1) R_1 = 390Ω and C_L = 50pF
   2) Active High R_1 = ∞; Active Low R_1 = 390Ω C_L = 50pF
   3) Active High R_1 = ∞; Active Low R_1 = 390Ω C_L = 5pF

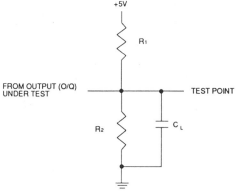

C_L INCLUDES JIG AND PROBE TOTAL CAPACITANCE

CAPACITANCE (T_A = 25°C, f = 1.0 MHz)

SYMBOL	PARAMETER	MAXIMUM*	UNITS	TEST CONDITIONS
C_I	Input Capacitance	8	pF	V_{CC} = 5.0V, V_I = 2.0V
$C_{I/O/Q}$	I/O/Q Capacitance	10	pF	V_{CC} = 5.0V, $V_{I/O/Q}$ = 2.0V

*Guaranteed but not 100% tested.

ELECTRICAL CHARACTERISTICS GAL16 / 20V8A-25L Commercial

Over Recommended Operating Conditions (Unless Otherwise Specified)

SYMBOL	PARAMETER	CONDITION	MIN.	TYP.	MAX.	UNITS
V_{OL}	Output Low Voltage		—	—	0.5	V
V_{OH}	Output High Voltage		2.4	—	—	V
I_{IL}, I_{IH}	Input Leakage Current		—	—	±10	µA
$I_{I/O/Q}$	Bidirectional Pin Leakage Current		—	—	±10	µA
I_{OS}[1]	Output Short Circuit Current	$V_{CC} = 5V$ $V_{OUT} = Gnd$	−30	—	−150	mA
I_{CC}	Operating Power Supply Current	$V_{IL} = 0.5V$ $V_{IH} = 3.0V$ $f_{toggle} = 15MHz$	—	75	90	mA

1) One output at a time for a maximum duration of one second.

DC RECOMMENDED OPERATING CONDITIONS GAL16 / 20V8A-25L Commercial

SYMBOL	PARAMETER	MIN.	MAX.	UNITS
T_A	Ambient Temperature	0	75	°C
V_{CC}	Supply Voltage	4.75	5.25	V
V_{IL}	Input Low Voltage	$Vss - 0.5$	0.8	V
V_{IH}	Input High Voltage	2.0	$Vcc+1$	V
I_{OL}	Low Level Output Current	—	24	mA
I_{OH}	High Level Output Current	—	−3.2	mA

SWITCHING CHARACTERISTICS — GAL16 / 20V8A-25L Commercial

Over Recommended Operating Conditions

PARAMETER	#	FROM	TO	DESCRIPTION	TEST COND.[1]	MIN.	MAX.	UNITS
t_{pd}	1	I, I/O	O	Combinational Propagation Delay	1	3	25	ns
	2	CLK	Q	Clock to Output Delay	1	2	15	ns
t_{en}	3	I, I/O	O	Output Enable, $Z \rightarrow O$	2	—	25	ns
	4	$\overline{OE}$	Q	Output Register Enable, $Z \rightarrow Q$	2	—	20	ns
t_{dis}	5	I, I/O	O	Output Disable, $O \rightarrow Z$	3	—	25	ns
	6	$\overline{OE}$	Q	Output Register Disable, $Q \rightarrow Z$	3	—	20	ns

1) Refer to **Switching Test Conditions** section.

AC RECOMMENDED OPERATING CONDITIONS — GAL16 / 20V8A-25L Commercial

PARAMETER	#	DESCRIPTION	TEST COND.	MIN.	MAX.	UNITS
f_{clk}	7	Clock Frequency without Feedback	1	0	33.3	MHz
	8	Clock Frequency with Feedback	1	0	28.5	MHz
t_{su}	9	Setup Time, Input or Feedback, before CLK ↑	—	20	—	ns
t_h	10	Hold Time, Input or Feedback, after CLK ↑	—	0	—	ns
t_w	11	Clock Pulse Duration, High	—	15	—	ns
	12	Clock Pulse Duration, Low	—	15	—	ns

SWITCHING WAVEFORMS

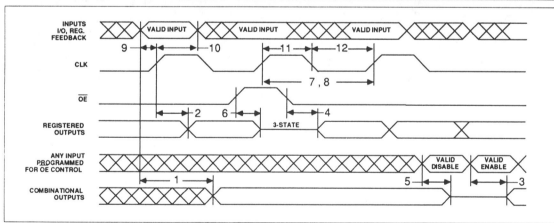